PICTURES
OF THE TROPICS

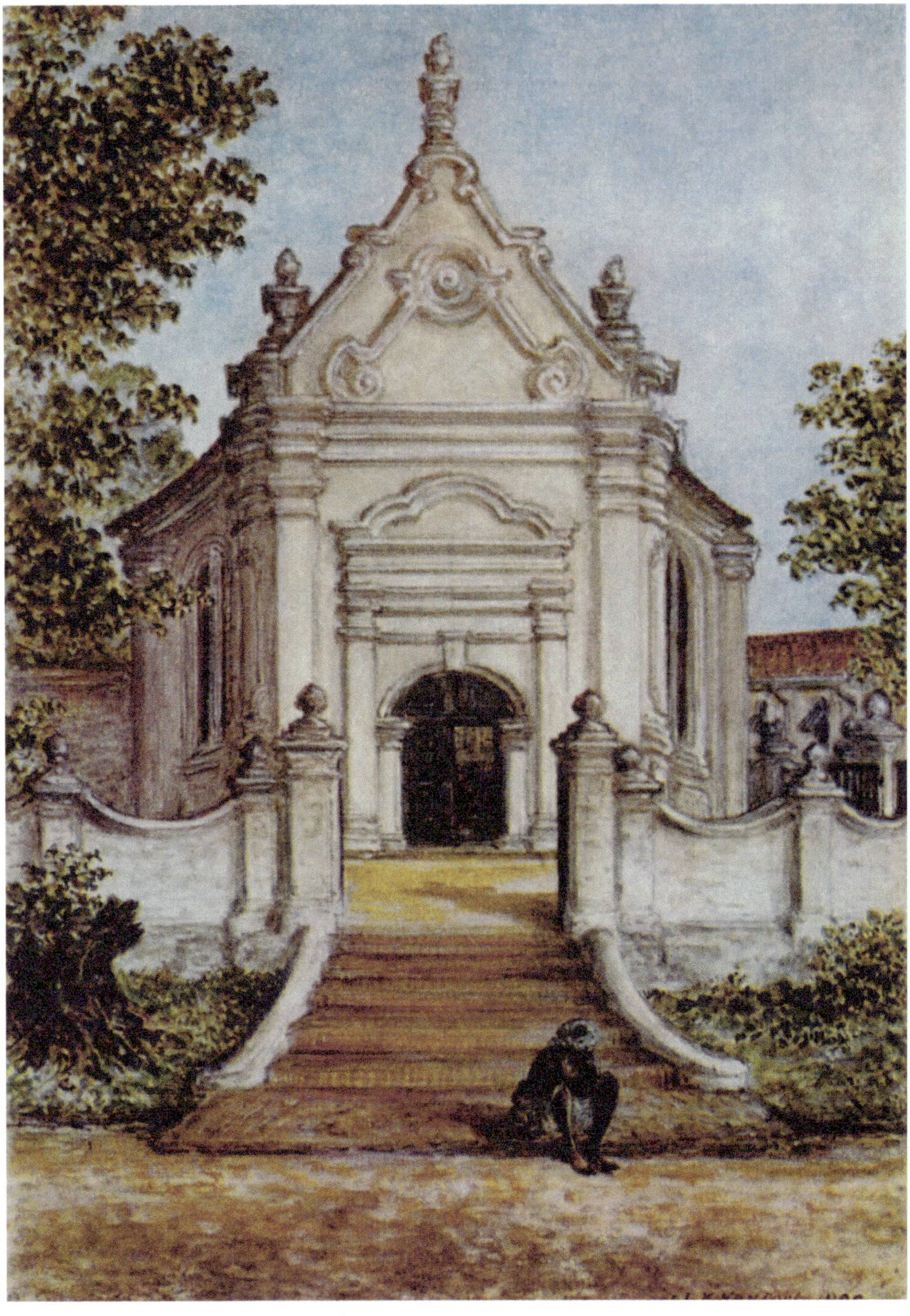

J. L. Kalenberg van Dort

Dutch Reformed Presbyterian Church, Galle O. 6, 25

KONINKLIJK INSTITUUT VOOR TAAL-, LAND- EN VOLKENKUNDE

PICTURES OF THE TROPICS

A CATALOGUE OF DRAWINGS,
WATER-COLOURS, PAINTINGS, AND SCULPTURES
IN THE COLLECTION OF THE
ROYAL INSTITUTE OF LINGUISTICS AND ANTHROPOLOGY
IN LEIDEN

BY

J. H. MARONIER

SPRINGER-SCIENCE+BUSINESS MEDIA, B.V. 1967

ISBN 978-94-017-6488-9 ISBN 978-94-017-6643-2 (eBook)
DOI 10.1007/978-94-017-6643-2

Softcover reprint of the hardcover 1st edition 1967

INTRODUCTION

During its existence of more than a hundred years the Royal Institute of Linguistics and Anthropology has seen a steady growth not only in its extensive library and collections of manuscripts and maps but also in its picture and photograph collections.

The picture collection comprises numbers of water-colours and drawings as well as original lithographs, etchings and woodcuts, some of which occur in illuminated manuscripts. The Institute also possesses a fair number of oil paintings and sculptures.

When surveying this collection of art it should be borne in mind that this owes its existence chiefly to a series of gifts made to the Institute in the course of the past century, rather than to an active policy of systematic acquisition, and that as a result it is a mixture of items of widely varying nature and artistic merit. Over the past few years, however, the Institute has had the opportunity to purchase important specimens to add to this collection. As the documentary value of the pictures was always the first consideration, the collection includes a number which from an artistic point of view might incur severe criticism. But by virtue of their subject-matter and dating these are interesting from an ethnographic point of view, the scenes and objects depicted in them having since disappeared or become totally lost as a result of change in the people's way of life.

The collection described in the present catalogue consists of objects of art executed in Indonesia, Surinam, the Netherlands Antilles and a number of other areas where the Dutch have settled at one time or another in the past. This geographical restriction is a consequence of the Institute's formerly having restricted its activities to these areas.

In cataloguing the above works a distinction has been made between the artistic expressions of the indigenous inhabitants of these areas and those of Dutch and other Europeans when confronted with landscapes, people and customs which were totally different from what they had been accustomed to in their own environment.

On the following pages a comprehensive summary will be given of all items of both categories which the Institute possesses, with the exception of the Oriental manuscripts. A detailed description will be given of each item, together with details of its provenance and its creator (as far as this is possible) and with references to the appropriate literature. In the descriptions captions, dates, place names and other particulars noted on the pictures themselves have been copied in their original spelling and language, followed by an English translation where this was necessary. Where there is no caption accompanying the original we have only given a description in English.

Taking into account the distinction indicated above the items have been further classified on the basis of the geographical areas to which they relate, namely Indonesia, the West Indies, Ceylon, South Africa and Japan. Then follows a short section containing descriptions of sculptures, which in turn is followed, in conclusion, by a section dealing with the work of Indonesian artists. Within these sections the items have been listed in alphabetical order on the basis of the names of their artists.

European Artists

During the earliest years of the existence of the Institute, when photography was still in its infancy, expeditions to areas which had as yet to be fully explored usually numbered artists among their staff to record in a series of drawings or water-colours everything the leaders of the expedition considered of ethnological interest. The illustrations in *Verhandelingen over de Natuurlijke Geschiedenis der Nederlandsche Overzeesche Bezittingen* (Treatises on the Natural History of the Netherlands Overseas Territories) (1839—1844), for instance, came about in this way. This volume includes lithographs based on water-colours and drawings by Q. R. M. Ver Huell, A. J. and J. Th. Bik, G. van Raalten, P. van Oort, Von Henrici, T. C. Bruining and Tsjing Wang Ho. There are a number of other travel journals which are illustrated with lithographs from drawings made during the expedition. Drawings thus made were sent to Holland along with the ethnographic material collected so that, in the event of the report of the expedition being published in book form, a selection could be made for illustrative purposes. The book by Dr. C. A. L. M. Schwaner, *Borneo: Beschrijving van het stroomgebied van de Barito* (Borneo: A description of the Barito River Area) (1853), one of the Institute's first publications, is an example of this. The water-colours and drawings made during this expedition by the artists Von Gaffron and Van Pers as well as by the leader of the expedition, a selection from which was made to serve as illustrations, were the first items in the Institute's art collection. And E. B. H. von Rosenberg's *Der Malayische Archipel* (1878), the original drawings for which are also included in the collection, is another example.

As the plates in these books are the result of a process by which the originals were freely copied on lithographs by lithographers and then printed, important details have in many cases been omitted. Greater credit should therefore be given to the originals than is due to the reproductions.

When we consider the excellent equipment, such as still and movie cameras and tape recorders, which is available to members of modern expeditions and enables them to collect (sometimes unseen) a wealth of useful data in a relatively short time, and when we compare this with the infinitely more difficult conditions under which artists accompanying expeditions over a century ago had to make their drawings and water-colours of landscapes and population types, we cannot help having a deep sense of admiration for their achievements, which made it possible to bring quantities of material back from journeys of exploration. The fact that three of the members of the Committee for Natural Sciences who joined the expedition to New Guinea and surrounding areas in 1828 died of tropical diseases shows how extremely difficult these conditions must have been in such matters as resistance to disease.

Classifying these professional artists in a separate category, a second category of artists is formed by naval and military officers, officials of the Colonial Administration and the Judicial apparatus, and physicians and planters who, impressed by the splendour of nature in the tropics and the artistic achievements of the indigenous people, devoted their natural talent for drawing to the preservation on paper or canvas of everything that aroused their interest. In this category, too, there are samples which testify to a remarkable talent for drawing, and which must be considered of importance for the study of the country and people concerned. Water-colours and drawings included as illustrations in manuscripts in the Institute's collection, such as, for example, in Paravicini di Capelli's important manuscript describing a journey in the Cape Colony, also belong to this category.

The third category consists of works by professional artists who, impressed by stories of relatives and friends in the Orient and the West Indies travelled to the tropics, usually for a short stay, to depict tropical landscapes and scenes of native life on their canvases; this they did from a typically European angle. Although artists like Isaac Israëls, Bleckman and Bauer (to mention only a few) returned to their home country with numbers of fine paintings and excellent studies, these works, because of their impressionistic style and flamboyant colours, often lack the details which are of interest to the student of ethnography. The work of the painter W. O. J. Nieuwenkamp, among others, who has had a large number of drawings, containing a wealth of information, reproduced in various publications, is an exception to this. Works in the Institute's collection which belong to this category are those by Fahringer and Sayers.

Indonesian Artists

In the category of objects of purely autochthonous art which originate from Indonesia the Institute counts among its collection a small number of precious specimens of Javanese and Balinese book illumination. In other Oriental manuscripts among its possessions too, the usually superb handwriting of the text is frequently interspersed with small decorative motifs. The pages are often edged with richly ornamented gilt and coloured borders. The Javanese manuscripts of the Babad Kartasoera (Cod. or. 15), a book of large format bound in gilded red leather, offers a splendid and unique example of this. It marks the climax of

Indonesian decorative art in the collection. A description of these manuscripts will have to follow in a separate catalogue of Oriental manuscripts. Besides this there are a number of beautifully illuminated books and pictures. Of these, a 31 metres long water-colour painting depicting the state procession of the Sultan of Jogjakarta, held on the 1st of January and the 31st of August, deserves separate mention. Another fine example is offered by the excellent water-colour paintings of the various uniforms of the ceremonial troops of the Sultan of Jogjakarta accompanied by descriptions is superb handwriting. The artist responsible for these, as well as for binding them into a book, was also a Javanese. There is, moreover, an extensive collection of water-colour paintings and drawings of creeses, spears, knives and other weapons, sunshades, head-dresses, dress ornaments and *batik* patterns.

An investigation conducted by the Colonial Government in 1939, the aim of which was to discover to what extent art-teaching in Bali had been adapted to the Balinese culture, resulted in an extensive collection of Balinese children's drawings which is now also included in the Institute's collection.

In conclusion we should mention the numerous letters addressed to Regents which were decorated with rich ornamentation along their edges. In some of these European influence is discernible, as in some of the illuminations of Javanese manuscripts.

Although as observed at the beginning of this introduction the Institute's collection is a rather haphazard one and contains works of wide diversity, owing its existence to a series of gifts, it nevertheless includes more than one item of exceptional merit, not only from an ethnologic but also from an artistic point of view.

It is the hope of the author of the present catalogue that an examination of its contents may result in other important works of art in the same field as those described in it, which may be present in the deposits of print galleries or museums or in private collections, being brought to the notice of the Institute in order to promote the study of this fields of decorative art.

In conclusion I wish to thank Dr. H. J. de Graaf, Dr. J. Noorduyn, Dr. C. Ouwehand, Dr. Th. P. Th. Pigeaud and Dr. P. Voorhoeve for their advice, of which grateful use has been made in the compilation of this catalogue. Sincere thanks are also extended to Mrs. M. C. J. Voorhoeve and to Miss M. J. L. van Yperen, who willingly undertook the burdensome task of turning the Dutch MS. of this book into English and have finalised the work in such a conscientious manner; to Mr. F. G. Jaquet for typing the MS and to Miss M. Winkel for reading the proofs.

CONTENTS

LIST OF PLATES

SELECTED BIBLIOGRAPHY

G. Brom, *Java in onze kunst,* Rotterdam 1931.

J. C. van Eerde, Javaansche dansen en Europeesche schilderijen. *Gedenkschrift Koninklijk Instituut,* 's-Gravenhage 1926, pp. 75-83.

P. J. C. Gabriël, Hollandsche schilderkunst in Indië. *Weekblad voor Indië,* X, 1913, pp. 266-268, 321-323, 345-347.

H. F. C. ten Kate, Schilder-teekenaars in Nederlandsch Oost- en West-Indië en hun beteekenis voor de land- en volkenkunde. *B.K.I.* LXVII, 1913, pp. 441-515.

J. de Loos-Haaxman, *Verlaat Rapport Indië.* Drie eeuwen Westerse schilders, tekenaars, grafici, zilversmeden en kunstnijveren in Nederlands-Indië, 's-Gravenhage (to be published 1968).

J. H. Maronier, Hollandsche teekenaars in Indonesië vóór honderd jaar. *Historia* XI, 1946, pp. 92-96.

P. A. J. Moojen, De schilderkunst. *Gedenkboek voor Nederlandsch-Indië t.g.v. het regeeringsjubileum . . . 1898-1923,* Batavia/Weltevreden/Leiden 1923, pp. 271-275.

K. Niehaus, Walter Spies en zijn werk. *Cultureel Indië* I, 1939, pp. 166-180.

K. Niehaus, Nog iets over leven en werk van Walter Spies. *Cultureel Indië* III, 1941, pp. 25-35.

G. P. Rouffaer, Voorrede in: W. O. J. Nieuwenkamp, *Bali en Lombok,* Edam 1906-1910, pp. I-VII.

G. P. Rouffaer, A. J. en J. Th. Bik. *Encyclopaedie van Nederlandsch-Indië,* 's-Gravenhage/Leiden 1917, Vol. I, pp. 302-303.

Walter Spies, *Schönheit und Reichtum des Lebens Walter Spies, Maler und Musiker auf Bali, 1895-1942.* Eine Autobiographie in Briefen mit ergänzenden Erinnerungen ges. und hrsg. von Hans Rhodius, Den Haag 1964.

J. Tielrooy, Indië in de schilder- en teekenkunst. *Elsevier's Geïllustreerd Maandschift* LXXX, 1930, pp. 1-10.

J. F. Stutterheim, *De teekeningen van Javaansche oudheden in het Rijksmuseum van Ethnographie,* Leiden 1933.

P. J. Veth, Javasche landschappen. *De Gids,* XXXVI, 1872, I, pp. 1-46.

P. J. Veth, *Catalogus der afdeeling Nederlandsche Koloniën van de Internationale en Uitvoerhandel Tentoonstelling te Amsterdam.* Leiden 1883, Groep I pp. 52-68, Groep II p. 9-11, 289-291.

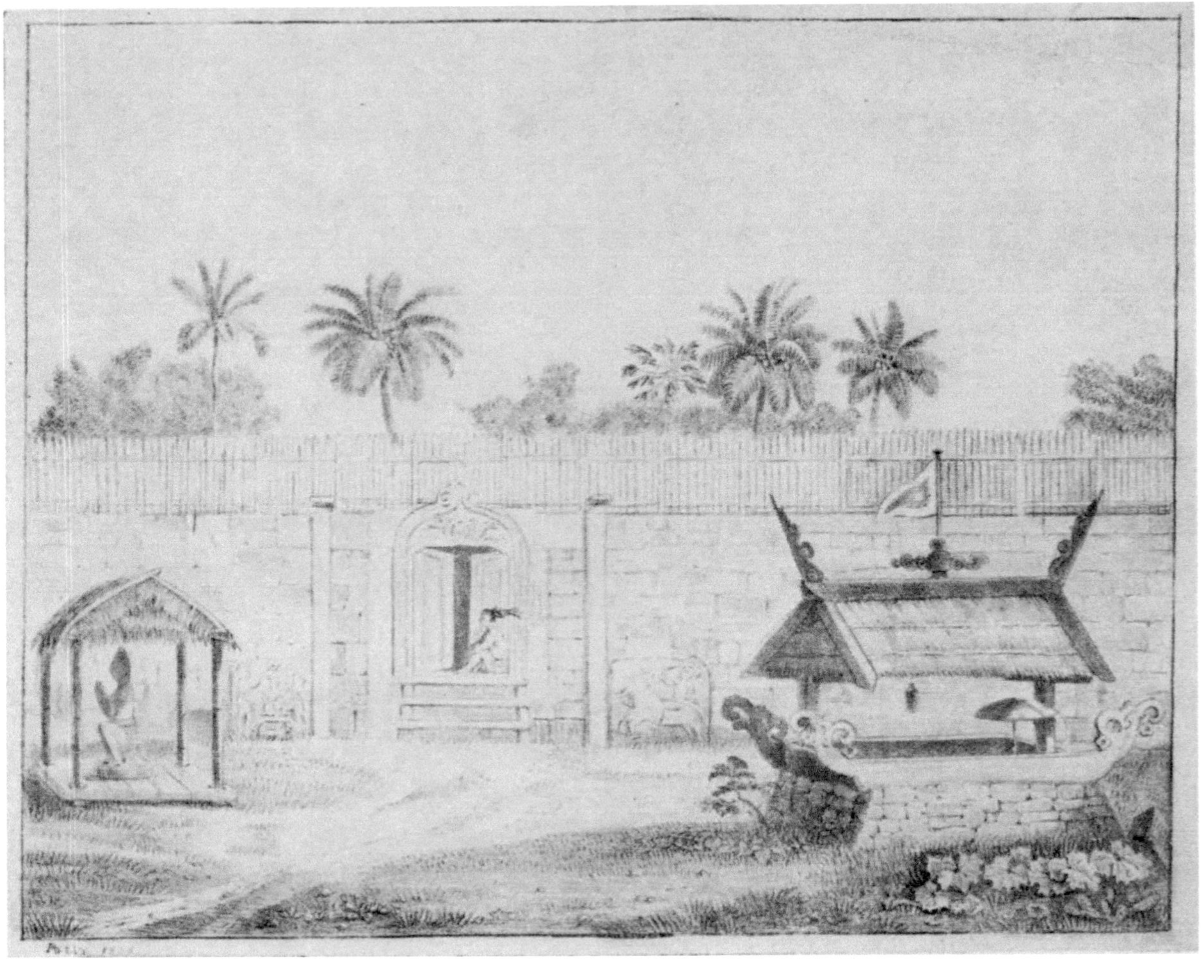

A. J. Bik Dula on the isle of Key F 14, 16

INDONESIA

BIK, Adrianus Johannes

Born Jan. 13, 1790, Duinkerken. Died Oct. 10, 1872, Brussels. Son of Jan Bik, merchant in Amsterdam. Went to the Indonesian Archipelago (1815). Was a draughtsman on the scientific expedition of Prof. Reinwardt (1821). Afterwards he made several journeys by order of the Government. Joined the Civil Service in 1828 as ass. resident of the surrounding districts of Batavia and was Bailiff of Batavia at the same time. Later he became a member of the Court of Justice. He returned to the Netherlands in 1847.

F. 14, 15 *House of the imam of Tual.*
Or. lith. Signed AB fec. 27 × 34,5 cm.
Note: Cf. A. J. Bik, *Dagverhaal eener reis, gedaan in het jaar 1824 tot nadere verkenning der eilanden Kefing, Goram, Groot en Klein Kei en de Aroe eilanden.* Uitgegeven door Mr. A. J. E. A. Bik. Leiden 1928 (Journal of a voyage made in the year 1824, for further reconnaissance of the islands of Kefing, Goram, Greater and Smaller Kei and the Aru islands. Published by Mr. A. J. E. A. Bik. Leiden 1928) p. 102—103.

F. 14, 16 *Gezigt op de poort van de met koraalsteen bemuurde negorij Doela op het eiland Kay en het Hindoes Beeld Déwa, benevens het graf van den radja van Doela* (View of the gate in the coral wall of the village of Dula, on the isle of Key. Hindu image Déwa and grave of rajah of Dula).
Or. lith. Signed AB lit 1828. 28,5 × 40 cm.
Note: Described in A. J. Bik, Journal etc., p. 100. The drawing for this lith. was made May 11th, 1824 (p. 113).

F. 14, 17 *Afgod en graf op Key* (Idol and grave on Key island).
Or. lith. Signed Bik. 16 × 24,5 cm.
Note: Detail of F. 14, 16.

BONNAT, Léon Joseph Florentin

Born June 20, 1833, Bayonne. Died Sept. 8, 1922, Monchy-St. Eloi. Painter of landscapes, historical scenes and especially of portraits. Pupil of Madrazzo, Madrid and L. Coquet, Paris. Was in Italy for 4 years and afterwards had a studio in Paris where Henri de Toulouse-Lautrec was one of his pupils. The "Musée Bonnat" was his gift to his native town.

S. 1 *Portrait of Isaac Dignus Fransen van de Putte* (1822—1902). Minister for the Colonies from 1863—1866. Patron of the Royal Institute.

Oil. Signed below on the left Ln Bonnat 1879. 144,5 × 105 cm. Gilt frame.

Provenance: Bequest of dr. J. H. Franssen van de Putte of The Hague to the National Museum Amsterdam (1907). Cat. 1920 No 558a. Transferred by the National Museum to the Service of the State's dispersed Objects of Art (inv. No. B 298); on loan to the Royal Institute.

BOUDEWIJNSE, A.

Born July 22, 1862, The Hague. Died Febr. 1909, The Hague. Pupil of the School of Arts in The Hague and of Cormon in Paris. Did much of his painting in Zealand, made many portraits.

S. 2 *G. H. van Soest,* joint founder of Indisch Genootschap (Indies Society).
Oil. Signed A. Boudewijnse 1887. 61 × 50 cm. Oak frame.

S. 3 *Th. F. Schill,* ex-member of the Factory of the Neth. Trading Co. Batavia.
Chalk. Signed A. Boudewijnse 1888. 57 × 40 cm. Oval, mounted, square black frame.

BREMER, F. C. A.

Art mistress at the Carpentier Alting School, a private secondary school and boarding school for girls at Weltevreden (Batavia).
Note: The dates with "Indonesië" probably added by the bookseller Loose.

F. 14, 1 *Portrait of Melchior Treub,* Director of "'s-Lands Plantentuin" (Governmental Botanic Gardens) at Buitenzorg (Bogor).
Black chalk. Dated [Indonesië 1909] (prior to Oct. 4, 1909 when Treub left for Europe). 51 × 35 cm.

F. 14, 2 *Portrait of Melchior Treub.*
Black chalk. Dated [Indonesië 1909] (prior to Oct. 4, see above). 48,5 × 30,5 cm.

F. 14, 3 *'Kebon' (gardener) with watering-cans.*
Black chalk. Dated [Indonesië 1909]. 50 × 31,5 cm.

F. 14, 4 *Chinese man sitting at a table.*
Black chalk on grey paper. Dated [Indonesië 1909]. 49 × 32 cm.

F. 14, 5 *Portrait of a European in the tropics.*
Black chalk. Signed FB [F. Bremer f. in Indonesië]. Dated 11-11-'09. 50 × 31 cm.

F. 14, 6 *Chinese head.*
Black chalk on brown paper. Signed FB [F. Bremer f. Indonesië]. Dated 21 april '09. 49 × 32,5 cm.

F. 14, 7 *Head of an old Indonesian woman.*
Red chalk on cream coloured paper. Signed FB [F. Bremer f. Indonesië]. Dated 8-8-'09. 50 × 31,5 cm.

F. 14, 8 *Portrait of a Hadji* (Arab?).
Black chalk on brown paper (left uppercorner torn off). Signed FB [F. Bremer]. Dated 5-8-'09. 34,5 × 22 cm.

F. 14, 9 *River with clumps of bamboo.*
Black chalk. Signed [F. Bremer fec. Indonesië 1908]. 34,5 × 27 cm.

Provenance: Purchased from the Antiquary Loose, The Hague, 1963.

CLERCQ, J. H. W. le

Born 1809. Died 1885, The Hague.
Was appointed full Lieutenant to the Major General of the Dutch Army F. B. Freiherr von Gagern (1794—1848) who was charged by Royal Decree, dated March 4, 1844, La X9 with the investigation of the defence system of the Neth. Indies. During this official tour from 1844—1846 of Java and Sumatra, and afterwards during a journey through India and Ceylon, le Clercq made many drawings and water-colour paintings. Cf. the biography of von Gagern: *Het leven van den Generaal Frederik von Gagern van 1813—1848 in Nederlandsche dienst,* naar het Hoogduits van H. von Gagern, door C. C. E. d'Engelbronner, 2 dln., Amsterdam 1866.

F. 2, 1—73 *Indonesia.*
Note: Arranged as far as possible in chronological order.

Batavia (Dec. 2, 1844—Febr. 28, 1845).

F. 2, 1 Page of sketchbook: *European house, Native hut, Chinese man, Native.*
Water-colours. Dated 15 Dec. 1844. 18 × 25 cm.

F. 2, 2 Page with sketches: *Palms, Natives and a Chinese man.*
Water-colours and ink. Dated Febr. 1845. Captions: klapper of kokospalm (coconut palm), Siwalangpalm, Areka, Bloem van de kokospalm (flower of the coconut palm). 17,5 × 25,5 cm.

Journey to Bantam (Febr. 28—March 11, 1845).

F. 2, 3 (Part of) *Minaret and Mosque at Bantam.*
Water-colours. Dated '2 Maart 1845'. 8,5 × 16 cm.

F. 2, 4 *In het Bantamsche* (Bantam landscape).
Water-colours. Dated 'Maart 1845'. 22,5 × 17 cm.
Note: The drawing shows a banyan tree with aerial roots, between which some native figures. This scene is mentioned in the biography, II, p. 287.

Journey through the Preanger-Regencies. March 14—27, 1845.
Cheribon-Samarang. March 27—April 4, 1845.

F. 2, 5 *Bij Kendang Badak* (Near Kendang Badak).
Ink and pencil touched up with water-colours. Dated '15 Maart 1845'. 17 × 19,5 cm.

F. 2, 6 *Kendang Badak.*
Ink and water-colours. Dated '15 Maart 1845'. 17 × 23,5 cm.
Note: The drawing shows a mountain landscape with bivouac and native figures. With notes in ink and pencil.

F. 2, 7 *Bivouac te Kendang badak* (Bivouac at Kendang badak).
Pencil. 14 × 16,5 cm.
F. 2, 8 *Top van de vulkaan Gedeh* (Peak of the vulcano Gedeh).
Water-colours. Dated '16 Maart 1845'. 8 × 17,5 cm.
Note: The view of the crater of Mt. Gedeh is described in the biography, II, p. 294.

F. 2, 9 *Op de Pangerango* (On Mount Pangerango).
Water-colours. Dated '16 Maart 1845'. 21 × 14 cm.
Note: Bears handwritten comment (excerpt from a letter?).

F. 2, 10 *Krater van de Telaga bodas* (Crater of Mt. Telaga bodas).
Water-colours. Dated '25 Maart 1845'. 12,5 × 12 cm.
Note: Mentioned in the biography, II, p. 297.

F. 2, 11 *Fumarole in de kraterwand van de Telaga bodas* (Fumarole in the wall of the crater of Mt. Telaga bodas).
Water-colours. Dated '25 Maart 1845'. 12,5 × 11 cm.

F. 2, 12 *Lake of Pandjalu* (?)
Water-colours. 9 × 15 cm.
Note: Probably this is the lake of Pandjalu, mentioned in the biography, II, p. 299. If this is so the drawing must have been made on March 29, 1845. On the reverse side are pencil notes of diary bearing the date 30 April [1845].

Journey through Kedu, Banjumas and Bagelen (April 25—June 3, 1845).

F. 2, 13 *Banteng Matie.*
Water-colours. Dated '14 Mei 1845'. 8,5 × 15,5 cm.

F. 2, 14 *Karangbolong* (?)
Pencil and water-colours. 9,5 × 15,5 cm.
Note: On reverse side: sea view with covered landing-place, with the caption: Karang ballang. 15 mei [1845].

F. 2, 15 *Bas-reliefs Tempel van Borabodor* (Bas-reliefs of the Temple of Borobudur).
Pencil. Dated '15 Maart 1845'. 17 × 25,5 cm.

F. 2, 16 *Javaansche graven* (Javanese graves).
Water-colours. 14 × 16,5 cm.

F. 2, 17 *Bewerking van sawah-velden met den patjol (hakijzer)* (Tilling the ricefields with the patjol (mattock).
Pencil. 17,5 × 24 cm.

F. 2, 18 *Bewerking van Sawah (Rijst-) velden* (Tilling Sawahs (Rice-fields)).
Pencil. 17,5 × 23 cm.

F. 2, 19 *Rijstcultuur (Sawas)* (Ricefields).
Ink and water-colours. 16 × 20,5 cm.

F. 2, 20 *Gamelan-orchestra.*
Water-colours. 18 × 26 cm.

F. 2, 21 *Tjandi Bima on the Diëng Plateau.*
Water-colours. Dated 'Diëng 3 Mei 1845'. 17 × 22 cm.
Note: Published in *B.K.I.* CIX, 1953, facing p. 228. The gamelan orchestra and group of figures to the right of the tjandi Bima also features in the lithographic print of a wajang-kulit performance, facing p. 8 of the 2nd Vol. of the *Tijdschrift voor Neêrl. Indië,* 8e jaargang, 1846.

F. 2, 22 *Rivier van Sindeh* (Sindeh River).
Water-colours. Dated '13 Mei 1845'. 22 × 14,5 cm.

F. 2, 23 *Offerande bij Merden* (Sacrificial near Merden).
Pencil. Dated 'Mei' [1845]. 17,5 × 22,5 cm.

Journey through the "Vorstenlanden" (Principalities) (Second half of June 1845).

F. 2, 24 *Top van de Merapi (gezien van Sello)* (Peak of Mt. Merapi (as seen from Sello)).
Water-colours. Dated '22 Junij 1845'. 11,5 × 19,5 cm.
Note: By Sello is meant the Sela on the saddle between Merbabu and Merapi, where there was a resthouse (pasanggrahan) of the Susuhunan.

F. 2, 25 *Brambanan* (Prambanan).
Water-colours. Dated '24 Junij 1845'. 11 × 12,5 cm.
Note: Vishnu temple?

F. 2, 26 *Brambanan* (Prambanan).
Water-colours. Dated '24 Junij 1845'. 14 × 12 cm.
Note: Small temple of the Lara Djonggrang-group? Published in *B.K.I.* CIX, 1953, facing p. 229.

F. 2, 27 *Tempel aan de voet van de Merapi* (Temple at the foot of Mt. Merapi).
Ink. Dated '31 Mei 1845'. 9,5 × 12 cm.
Note: Tjandi Asu. Published in *B.K.I.* CIX, 1953, facing p. 229.

East-Java (Aug. 23—Nov. 2, 1845).

F. 2, 28 *Residentiehuis van Soerabaja te Simpang* (Mansion of Resident of Surabaja at Simpang).
Pencil. Dated 1846. 15 × 19 cm.

F. 2, 29 *Image of watchman near the aloon-aloon at Singasari, Malang.*
Pencil. 7,5 × 8 cm.

F. 2, 30 *A European, sitting in a chair* (?).
Pencil. 5,5 × 5 cm.

F. 2, 31 *Nandi.*
Pencil. 3,5 × 6 cm.
Note: Nos. F. 2, 29—31 are pasted onto 1 sheet, a small note has been pasted on the same sheet with the comment: Deze teekeningen zijn op reis gemaakt door den Hoog Edel Gestrengen Heer Generaal Majoor I. H. W. Le Clercq (sic), in different handwriting: geb. 1809, overl. 1885 's-Gravenhage (These drawings were made during a voyage by the Right Honourable Major General I. H. W. Le Clercq: born 1809, died 1885 The Hague).

F. 2, 32 *Penataran.*
Water-colours. Dated 11 October 1848. 18 × 23 cm.
Note: Temple of Dates and Principal temple. Published *B.K.I.* CIX, 1953, facing p. 228.

Voyage from Padang to Tapanoeli (June 12 and first half of July 1845).

F. 2, 33 *Bij Padang* (Near Padang).
Water-colours. Dated '1 Junij 1846'. 15,5 × 23,5 cm.

F. 2, 34 *Sumatra Laras huizen Singkarah* (Sumatra Laras houses Singkarak).
Water-colours. Dated '14 Junij 1846'. 9,5 × 15 cm.
Note: In pencil faintly legible: Huis van Dato Bandoro (?) (House of Dato Bandoro).

F. 2, 35 *Karbouwengat.*
Water-colours. Dated [about the middle of June] 1846. 9 × 12,5 cm.

F. 2, 36 *Loender.*
Water-colours. Dated '1 Julij 1846'. 9,5 × 16,5 cm.

F. 2, 37 *Pionghaij.*
Water-colours. Dated '4 Julij 1846'. 10,5 × 14,5 cm.
Note: Probably Li Piongot.

F. 2, 38 *Begraafplaats der Radja's (Batta-landen) Siaboe* (Chiefs' burial place (Batak-country) Siabu).
Water-colours. Dated '9 Julij 1846'. 17 × 25 cm.

F. 2, 39 *Bivouac op den Loeboe radja* (Bivouac on the Lubuk Raya).
Water-colours. Dated '14 Julij 1846'. 10,5 × 16 cm.

F. 2, 40 *Loeboe radja.*
Water-colours. Dated '14 Julij' [1846]. 16 × 10,5 cm.
Note: The drawing shows a camp-fire at Lubuk Raya.

F. 2, 41 *Landscape with road.*
Water-colours. Dated [July, 1846]. 9,5 × 14,5 cm.

F. 2, 42 *Sumatra. Hangende rottang-brug* (Sumatra. Rattan suspension bridge).
Water-colours. Dated 1846. 20,5 × 27,5 cm.

Voyage from Benkoelen to Palembang (Second half of Aug., 1846).

F. 2, 43 *Benkoelen. Regent met gevolg* (Benkulen. Regent with retinue).
Water-colours. 25 × 19 cm.

F. 2, 44 *Camp-fire.*
Ink and pencil. 10 × 13 cm.

F. 2, 45 *Native men with lances around a camp-fire.*
Ink and pencil. Dated (partly effaced) 16 Aug. 1846. 12 × 10,5 cm.

F. 2, 46 *Camp.*
Ink and pencil. Dated (partly effaced) Penamben (Penamping?) 17 Aug. 1846.
Note: Nos. F. 2, 44—46 are pasted onto the same sheet.

F. 2, 48 *Barang di Danau. Bivouac.*
Water-colours. Dated '20 Augustus 1846'. 19,5 × 25,5 cm.

F. 2, 49 *Priegie.*
Water-colours. Dated '22 Augustus 1846'. 15 × 12 cm.

F. 2, 50 *Camp-fire with natives and sentry.*
Ink and pencil. 10,5 × 14,5 cm.

F. 2, 51 *Construction of a temporary bridge.*
Ink and pencil. 10,5 × 14 cm.
Note: Nos. F. 2, 50 and 51 are pasted onto the same sheet.

Journey home via Ceylon, India, Egypt (Nov. 1, 1846— June 30, 1847).

F. 2, 52 *Poeloe Pinang.*
Water-colours. Dated 10 Nov. 1845 (read: 1846). 17 × 22 cm.

F. 2, 53 *Aren-palm* (Sugar palm).
Water-colours. 20 × 16,5 cm.
Note: Same technique and on the same paper as F. 2, 52.

F. 2, 47 *Native houses and villages in India* (?).
Ink and pencil. 8,5 × 15,5 cm.
Note: On the same sheet as F. 2, 44—46.

Native types, mostly from Sumatra.

F. 2, 54 *Agam: Gala-tenue der Hoofden.* (Agam: Chiefs' Gala-dress). Water-colours. Dated 1845 in pencil. 22,5 × 17,5 cm.

F. 2, 55 *Agam: Gala-tenue vrouwen der Hoofden* (Agam: Gala-dress of Chiefs' wives).
Water-colours. Dated 1845 in pencil. 23 × 17,5 cm.

F. 2, 56 *Paija Combo.*
Water-colours. Dated 1845 in pencil. 23 × 18 cm.
Note: The drawing shows women from Pajakumbo.

F. 2, 57 *XIII Kota's. Hoofd in Gala-kleeding* (XIII Kotas. Chief in Gala-dress).
Water-colours. Dated 1845 in pencil 22 × 17 cm.

F. 2, 58 *XIII Kota's.*
Water-colours. Dated 1845 in pencil. 24 × 17,5 cm.
Note: The drawing shows an armed native (woman in background). On reverse side pencil sketches of both figures.

F. 2, 59 *Lima-poeloe. Toewanko Gedang.*
Water-colours. Dated 1845 in pencil. 24,5 × 19 cm.
Note: The drawing shows Tuwanku Gedang on horse-back.

F. 2, 60 *Lima-poeloe. Toewanko Gedang.*
Water-colours. Dated 1845 in pencil. 24 × 19,5 cm.
Note: The drawing shows Toeanko Gedang, walking.

F. 2, 61 *Lima-poeloe. Vrouw van een Hoofd* (Lima-pulu. Wife of a Chief).
Water-colours. Dated 1845 in pencil. 22 × 16 cm.

F. 2, 62 *XIII Kota's.*
Water-colours. Dated 1845 in pencil. 25 × 19,5 cm.
Note: The drawing shows a native with 2 women carrying water.

F. 2, 63 *Lima Poeloe.*
Water-colours. Dated 1845 in pencil. 22 × 17 cm.
Note: The drawing shows an armed native and 2 men in background.

F. 2, 64 *Javanese chief with servant carrying payong.*
Water-colours. Dated 1845 in pencil. 23 × 20,5 cm.

Copies

F. 2, 65 *Koetsier* (Coachman (in Batavia?)).
Water-colours. Note: Kopien naar teekeningen van Hardouin (Copies based on drawings by Hardouin). Dated 1845. 17 × 11 cm.
Note: Nos. F. 2,65—70, and possibly 71, appear to be based on unknown drawings by Ernest Hardouin.

F. 2, 66 *Mahomed. Priester* (Mahomed(an) Priest).
Water-colours. Note: 'Idem'. Dated 1845. 24,5 × 17,5 cm.

F. 2, 67 *Javaansch Hoofd* (Javanese Chief).
Water-colours. Note: 'Idem'. Dated 1845. 24,5 × 17,5 cm.

F. 2, 68 *Native woman after bathing in river.*
Water-colours. Note: 'Idem'. Dated 1845. 25 × 17,5 cm.

F. 2, 69 *Maid-servant on the front verandah.*
Water-colours. Note: 'Idem'. Dated 1845. 24 × 17,5 cm.

F. 2, 70 *Fisherman with casting-net.*
Water-colours. Note: 'Idem'. Dated 1845. 25 × 17,5 cm.

F. 2, 71 *Waterdrager Batavia* (Water-carrier (in) Batavia).
Water-colours. Dated 1845. 17 × 14 cm.

F. 2, 72 *View of the jungle near the Tji-Tarum River.*
Water-colours. Dated in pencil 1846. 20,5 × 27 cm.
Note: Copied from lithograph in: C. L. Blume, *Rumphia sive Commentationes Botanicae...* 2 Vols. Lugduni-Batavorum 1835—1836, Tab. 38: Nemus intactum e regione Artocarpearum circa fluvium Tji-Tarum, drawn from nature by A. A. J. Payen.
Caption: the word *copie* (copy) and the names of the trees, as in the lithograph.

F. 2, 73 *View of rocky coast with threatening gale.*
Water-colours. Dated in pencil 1846. 20,5 × 28 cm.
Note: Copied from Tab. 121 (*ibidem*): Prospectus orae litoreae praeruptae et desertae palmis rari consitae, oreinte procella, painted by Raden Saleh. Same captions.

India (F 3, 1—86)

F. 3, 1 *Madras.*
Water-colours. Dated 'December 1846'. 17,5 × 25 cm.
Note: View of Madras from the sea.

F. 3, 2 *Calcutta.*
Water-colours. Dated '29 December 1846'. 17,5 × 25,5 cm.
Note: Palanquin in front of Mosque.

F. 3, 3—6 *Calcutta.*
Note: 4 studies on one sheet.

F. 3, 3 *Indian (native?) women.*
Ink, pencil and water-colours. 9,5 × 11,9 cm.

F. 3, 4 *Bedienden Calcutta* (Servants Calcutta).
Ink, pencil and water-colours. 9,5 × 15,5 cm.

F. 3, 5 *Types of Calcuttans.*
Ink, pencil and water-colours. 9,8 × 15,5 cm.

F. 3, 6 *Types of Calcuttans.*
Ink, pencil and water-colours. 9,4 × 16,3 cm.

F. 3, 7—8 *Calcutta.*
Note: 2 studies on one sheet.

F. 3, 7 *Palankin Calcutta* (Palanquin Calcutta).
Water-colours. 8,3 × 15,5 cm.
Note: Palanquin in front of house.

F. 3, 8 *Wagen van inlandsch fabrikaat* (Indian made carriage).
Water-colours. 9,5 × 15,9 cm.

F. 3, 9—10 *Calcutta and Simla.*
Note: 2 studies on one sheet.

F. 3, 9 *Riv. Hooglij nabij Calcutta* (Riv(er), Hooghly near Calcutta).
Pencil. 9,6 × 16,3 cm.
Note: River with sailing-boat.

F. 3, 10 *Simla.*
Pencil. 10,5 × 16,4 cm.

F. 3, 11 *Evening excursion in palanquin.*
Water-colours. Dated 'Januarij' [1847]. 17,6 × 25,5 cm.

F. 3, 12 *Benares.*
Water-colours. Dated 'Jany.' [1847]. 17,6 × 25,5 cm.
Note: The drawing shows a palace on the Ganges.

F. 3, 13 *Benares.*
Water-colours. Dated Jan. [1847]. 17,8 × 25,5 cm.
Note: The drawing shows temples on the Ganges, Ghats.

F. 3, 14—17 *Near Allahabad and Cawnpore.*
Note: 4 studies on one sheet.

F. 3, 14 *Bij Allahabad* (Near Allahabad).
Ink and pencil. Dated '12 Jany' [1847]. 9,5 × 13,5 cm.
Note: The drawing shows a well.

F. 3, 15 *Bij Allahabad* (Near Allahabad).
Ink and pencil. Dated '12 Jany' [1847]. 8,5 × 13,8 cm.
Note: The drawing shows a picnic.

F. 3, 16 *Ganges Riv. Caunpoor* (Cawnpore).
Ink and pencil. Dated 'Jany' [1847]. 8,8 × 16,3 cm.
Note: The drawing shows a houseboat.

F. 3, 17 *Ferry-boats.*
Ink and pencil. 8,8 × 13,3 cm.

F. 3, 18 *Residentiehuis te Luknow (Koningrijk Oude)* (Governor's house at Luknow (Kingdom of Oudh)).
Ink. On brown paper. 14,3 × 17 cm.

F. 3,19 *Bungalow.*
Water-colours. Dated '19 Jany 1846' (read: 1847). 17,5 × 22,7 cm.
Note: The drawing shows a palanquin with figures in front of house.

F. 3, 20 *Tombes tusschen Agra en Cawnpoor* (Tombs between Agra and Cawnpore).
Ink and pencil. Dated '19 Janij' [1847]. 9,6 × 12,9 cm.

F. 3, 21 *Memorial tomb.*
Ink and pencil. 9,2 × 15,6 cm.
Note: On one sheet with F. 3, 20.

F. 3, 22 *Agra.*
Ink and pencil. Dated 21 Jan. [1847]. 8,6 × 16,2 cm.
Note: The drawing shows a camp with camels.

F. 3, 23 *Cart.*
Water-colours. 9 × 15,5 cm.
Note: On one sheet with F. 3, 22 and 24.

F. 3, 24 *Oxen and waggons under a tree.*
Ink and pencil. 9,8 × 14,2 cm.
Note: On one sheet with F. 3, 22 and 23.

F. 3, 25 *Taj Mahal, Agra.*
Water-colours. Dated 22 'Januarij' [1847]. 17,5 × 25,5 cm.

F. 3, 26 and 27 *Agra and Ellora.*
Note: 2 studies on one sheet.

F. 3, 26 *Agra. Thatch Mahal* (Taj Mahal, Agra).
Ink. On blue paper. 11,8 × 14 cm.

F. 3, 27 *Roza Ellora.*
Ink. On blue paper. 12,5 × 12,6 cm.

F. 3, 28 *Futtipoor Sikrie* (Fatehpur-Sikri).
Water-colours. Dated '23 Januarij' [1847]. 16,2 × 25,5 cm.
Note: The drawing shows tents and carts in front of the Palace.

F. 3, 29 *Futtipoor Sikrie* (Fatehpur-Sikri).
Water-colours. Dated '23 Januarij' [1847]. 17,5 × 25,5 cm.
Note: The drawing shows the Palace.

F. 3, 30 *Bhurtpoor. Oud paleis van den Radja* (Bharatpur Former palace of the Rajah).
Water-colours. Dated '24 Jany' [1847]. 17,5 × 25,5 cm.

F. 3, 31 *Bhurtpoor. Bungalow van den Radja* (Bharatpur. Rajah's bungalow).
Water-colours. Dated '24 Jany' [1847]. 17,6 × 25,5 cm.

F. 3,32 *Paleis te Deeg* (Palace at Dig).
Water-colours. Dated '25 Januarij' [1847]. 17,5 × 25,5 cm.

F. 3, 33 *Paleis van den Radja van Bhurtpoor te Deeg* (Palace of the Rajah of Bharatpur at Dig).
Water-colours. Dated 25 Jan. [1847]. 17,5 × 25,5 cm.
Note: The drawing shows a hall with three European gentlemen dining

by candle-light. Probably Fr. von Gagern, van Aersen and the artist, J. H. W. le Clercq.

F. 3, 34 *Bindrabun* (Bindraban).
Water-colours. Dated '27 Jany' [1847]. 17,5 × 25,5 cm.

F. 3, 35 *Column of the palace of Bindraban.*
Water-colours. 13,2 × 17,2 cm.

F. 3, 36 *Kootub bij Delhie* (Qutb Minâr near Delhi).
Water-colours. Dated '31 Januarij' [1847]. 17,5 × 25,5 cm.

F. 3, 37 *Kootub bij Delhie* (Qutb Minâr near Delhi).
Water-colours. Dated '31 Januarij' [1847]. 25,5 × 17,5 cm.
F. 3, 38 *Colachita.*
Ink and pencil. Dated '2 Februarij' [1847]. 17,8 × 25,5 cm.

F. 3, 39 *Punjab.*
Ink and pencil. Dated '10 Februarij' [1847]. 17,5 × 25,5 cm.
Note: The drawing shows camels at a well.

F. 3, 40 *Lahore. Wandelrid van den Maharadja Djulup Singh* (Lahore. Maharajah Dalip Singh going for a ride).
Water-colours. Dated '11 Februarij 1846' (read: 1847). 19,4 ×22,5 cm.
Note: This refers to the young Maharajah Dalip Singh (still under age) who was dethroned on 29th March, 1849, after the second Sikh-war (1848—1849) and afterwards lived in England on a sum paid in compensation.

F. 3, 41 *Tombe bij Ferozeshah* (Memorial tomb near Ferozeshah).
Water-colours. Dated 1847. 22 × 17,5 cm.
Note: The three figures shown on the drawing are probably the same as in F. 3, 33. The inscription on the tomb as written in ink on the drawing: Sacred to the memorie of Lt. Peter Colnett Lambert, 3d troop 3d Brigade Horse Artillerie, who fell in action at Ferozeshah on the 22th Dec. 1845. Erected over his remains bij his brother officers.
The spot is lonelij and far awaij, but the brothers do not forget.
Note: During the first Sikh-war the battle near Ferozeshah (Firuzshuhur) was fought on 21—22 Dec. 1845.

F. 3, 42 *Lahore (Punjab) Salamar garden (tuin van de Koningin)* (Lahore (Punjab) Shâlamar garden (the Queen's garden)).
Water-colours. Dated '12 februarij 1846' (read: 1847). 20,4 × 25,5 cm.

F. 3, 43 *Hoofdtoisel der Afghanen. Lahore* (Afghan head-dress. Lahore).
Ink and pencil. 12,2 × 9,1 cm.
Note: The picture bears the note: dit zijn werpringen (these are quoits).

F. 3, 44 *IJzeren werpringen, rondom scherp, een wapen der Afghanen, eenige dier ringen zijn gedragen op den tulband* (Iron quoits, sharp-edged on the outside, an Afghan weapon. Some of these quoits are worn on top of the turban).

Ink and pencil. 15,5 × 20 cm.
Note: On one sheet with F. 3, 43.

F. 3, 45 *Umritsir* (Amritsar).
Ink and pencil, with grey wash. Dated '14 Februarij 1846' (read: 1847). 17,5 × 25,5 cm.

F. 3, 46 *Simla. (Hymalaija)* (Himalayas).
Ink. Dated '21 Februarij' [1847]. 17,8 × 25,5 cm.
Note: The drawing shows a mountain village.

F. 3, 47 *Simla. (Hymalaia)* (Himalayas).
Ink. Dated 21 Febr. [1847]. 17,5 × 25,5 cm.
Note: The drawing shows a mountain-side with houses.

F. 3, 48 *Himalaija* (Himalayas).
Water-colours. Dated 'Februarij 1846' (read: 1847). 17,2 × 22,8 cm.
Note: The drawing shows 3 riders crossing a mountain stream.

F. 3, 49 *Hijmalaija* (Himalayas).
Water-colours. Dated 'Februarij 1847'. 23 × 17,2 cm.
Note: The drawing shows a camp.

F. 3, 50 *Dolpoor* (Dholpur).
Water-colours. Dated '3 Maart 1846' (read: 1847). 16,8 × 23 cm.
Note: The drawing shows a state coach with six elephants.

F. 3, 51 *Balwarra.*
Water-colours. Dated '19 Maart 1846' (read: 1847). 17,5 × 25,5 cm.
Note: The drawing shows a barn with trees.

F. 3, 52 *Doorwau* (Khandwa?).
Water-colours. Dated '20 Maart 1846' (read: 1847). 17,5 × 25,5 cm.
Note: The drawing shows a roadside resthouse with three Europeans, probably the same as on F. 3, 33. With a note: (Dürmasaleh) 112° Fahrenheit. On reverse side sketch of fan-palm.

F. 3, 53 *Asschguhr* (Asirgarh).
Water-colours. Dated '21 Maart 1846' (effaced but faintly legible ...7). 15,7 × 19,2 cm.
Note: The drawing shows a fort on mountain top.

F. 3, 54 *Balwarra* (Barwaha?).
Water-colours. Dated '19 Maart 1846' (read: 1847). 25 × 17,5 cm.
Note: The drawing shows a landscape with tree in foreground.

F. 3, 55 *Ajunta-caves* (Ajanta caves).
Water-colours. Dated 'Maart 1846' (read: 1847). 17,5 × 25,5 cm.

F. 3, 56 *Ajunta-caves* (Ajanta caves).
Water-colours. Dated 'Maart (effaced but faintly legible:) 1847'. 17,5 × 25,5 cm.
Note: The drawing shows a temple with human figures (the artist with a guide?).

F. 3, 57 *Rosa-Ellora.*
Water-colours. Dated 'Maart' [1847]. 17,5 × 25,5 cm.
Note: The drawing shows a temple.

F. 3, 58 *Rosa-Ellora.*
Water-colours. Dated 'Maart' [1847]. 17,8 × 25,5 cm.
Note: The drawing shows the same temple as F. 3, 57, with 2 human figures in foreground.

F. 3, 59 *Rosa-ellora.*
Ink and pencil. Dated 'Maart 1846' (read: 1847). 17,5 × 25,5 cm.

F. 3, 60 *Poolmurrij.*
Water-colours. Dated '28 Maart 1846' (read: 1847). 17,5 × 25 cm.
Note: The drawing shows a rest-house. With a note 111° Fahrenheit.

F. 3, 61 *Elephanta bij Bombaij* (Elephanta near Bombay).
Water-colours. Dated April 1846 (read: 1847). 14 × 23 cm.

F. 3, 62 *Khandala.*
Ink and pencil. Dated 6 April 1846 (read: 1847). 17,5 × 25,5 cm.
Note: The drawing shows a lake and mountains.

F. 3, 63 *Sira.*
Water-colours. Dated '30 Mei 1847' (probably: 30 April 1847). 17,5 × 25,5 cm.
Note: The drawing shows ships in roads near Aden.

F. 3, 64 *Gewatteerde kleeding in Hindostan* (Padded garments in Hindustan).
Water-colours. 15,5 × 16,5 cm.
Note: The drawing shows a man with a camel in background.

F. 3, 65 *Djam. Sidien.*
Pencil. 15 × 11 cm.
Note: Portrait study; on one sheet with F. 3, 66.

F. 3, 66 *Indian woman* (with details of ear-ornament).
Water-colours. 8,8 × 9,9 cm.
Note: 'Oor met gouden ornementen; die ornementen hebben een stift, waarop een moertje wordt geschroefd' (Ear with golden ornaments; these ornaments are fitted wih a threaded wire onto which a tiny nut is screwed to fasten to ear).

F. 3, 67 *Halt at a well with a camel beside it.*
Ink. 14,5 × 16,3 cm.

F. 3, 68 *Militaire transporten. Hindostan* (Military transports, Hindustan).
Ink and pencil. 15 × 23 cm.
Note: The drawing shows a military transport with elephant and camels.

F. 3, 69 *Prayer niche near a well.*
Ink and water-colours. 9 × 15 cm.
Note: 3, 69 and 70 on one sheet.

F. 3, 70 *Tent under a tree.*
Ink and water-colours. 9 × 16,3 cm.

F. 3, 71 *Elephant.*
Ink and pencil. 12 × 9,5 cm.
Note: 3, 71 and 72 on one sheet.

F. 3, 72 *Accident during a lion hunt.*
Pencil. 10 × 18 cm.

F. 3, 73 *Camel.*
Ink and pencil. 9 × 9 cm.
Note: 3, 73—75 on one sheet.

F. 3, 74 *Camel.*
Ink and pencil. 8 × 13,3 cm.
Note: With note: een ijzeren ring gaat door het middelschot van den neus, waaraan een koord tot bestiering. Een lange aan de halster. (An iron ring is passed through the centre of the nose and a cord to guide the animal by attached to it. A long cord is attached to the halter.)

F. 3, 75 *Bepakking van een Olijphant met de hawda* (Elephant saddled with howdah).
Ink. 10 × 10,5 cm.

F. 3, 76—83 *Native types of Bombay, Calcutta, Hindustan.*
Ink, pencil and water-colours.
Note: Eight very small studies on one sheet, showing, among others, Parsees at Bombay; girls of Bombay; a fakir; Hindustan head-dress.

The following drawings were probably not drawn from nature.

F. 3, 84 *Baijaderes. Bengalen* (Dancing-girls, Bengal).
Ink. On brown paper. 20 × 13,5 cm.

F. 3, 85 *Engelsch Indie. Officieren der irreguliere Kavallerie* (British India. Officers of the irregular Cavalry).
Ink. On brown paper. 19,5 × 17,5 cm.

F. 3, 86 *Peshawur. Transport van kanons 9 lb over het gebergte* (Peshawar. Transport of 9 lb canons crossing the mountains).
Ink. On brown paper. 13 × 21 cm.

Ceylon (F. 3, 87—104).

F. 3, 87 *Point de Galle.*
Water-colours. Dated 17 Nov. 1846. 9,8 × 8,8 cm.

F. 3, 88 *Kandij. Paleis van de oude Koningen* (Kandy. Former Royal Palace).
Water-colours. Dated 19 Nov. 1846. 12,6 × 9,4 cm.

F. 3, 89 *Kandij.*
Water-colours. Dated 19 Nov. 9,8 × 8,7 cm.

F. 3, 90 *Ferry-boat.*
Water-colours. Dated 'Ceilon November 1846'. 25,5 × 17,8 cm.

F. 3, 91 *Street-seller.*
Water-colours. Dated 'Ceilon November 1846'. 19,9 × 14 cm.

F. 3, 92 *Smoking man in front of a house.*
Water-colours. Dated 'Ceilon Nov. 1846'. 16,6 × 11,4 cm.

F. 3, 93 *Man and woman with a house in background.*
Water-colours. Dated 'Ceilon November 1846'. 19,7 × 13,3 cm.

F. 3, 94 *Landscape with bridge.*
Water-colours. Dated 'Ceilon. Point de Galle November 1846'. 20,7 × 17,8 cm.

F. 3, 95 *Bij Point de galle* (Near Point de Galle).
Water-colours. Dated 2 December 1846. 25,5 × 17,6 cm.

F. 3, 96 *View of Point de Galle from the sea.*
Water-colours. Dated 'Ceilon. Point de Galle December 1846'. 24 × 15,8 cm.
Note: The drawing shows a view of Point de Galle from the sea.

F. 3, 97 *A fort.*
Water-colours. Dated 'Ceilon. Point de Galle December 1846'. 23,7 × 17,6 cm.

F. 3, 98 *Bij Point de Galle* (Near Point de Galle).
Water-colours. Dated 1846. 7,4 × 12,9 cm.

F. 3, 99 *Roads with sailing-ships.*
Water-colours. Dated 'Point de Galle 1846'. 14,9 × 13,7 cm.

F. 3, 100 *Bij Point de Galle* (Near Point de Galle).
Water-colours. On brown paper. Dated 1846. 25,8 × 17,8 cm.
Note: The drawing shows a view of a beach with palmtrees and boats.

F. 3, 101 *Cingalezen (Ceilon)* (Singhalese people (Ceylon)).
Ink. 11 × 15,6 cm.
Note: 3, 101—103 on one sheet.

F. 3, 102 *Boedha priesters (Ceilon)* (Buddhist priests (Ceylon)).
Ink. 6,4 × 11,3 cm.

F. 3, 103 *Portrait head.*
Pencil. 8,2 × 9,6 cm.

F. 3, 104 *Ceilon.*
Ink. On brown paper. 19,1 × 17,9 cm.
Note: The drawing shows an outrigged proa in heavy seas.

Provenance: F. 2 bought at the auction of Van Stockum's antiquary in The Hague on Nov. 19, 1952. Probably from the estate of the late Mr. N. Posthumus, second-hand bookseller in The Hague. F. 3 bought from Oriental Bookshop, The Hague 1965.

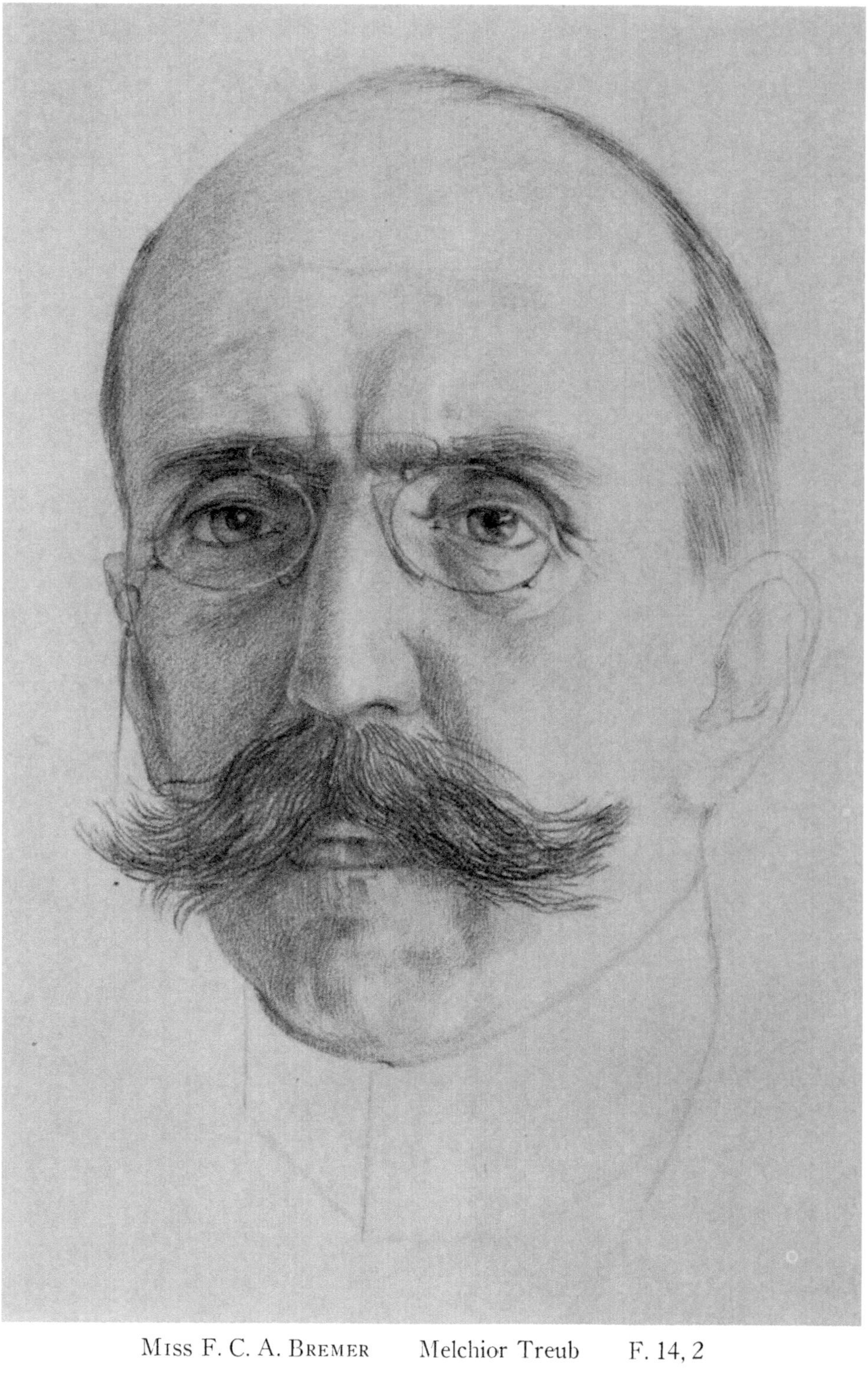

MISS F. C. A. BREMER Melchior Treub F. 14, 2

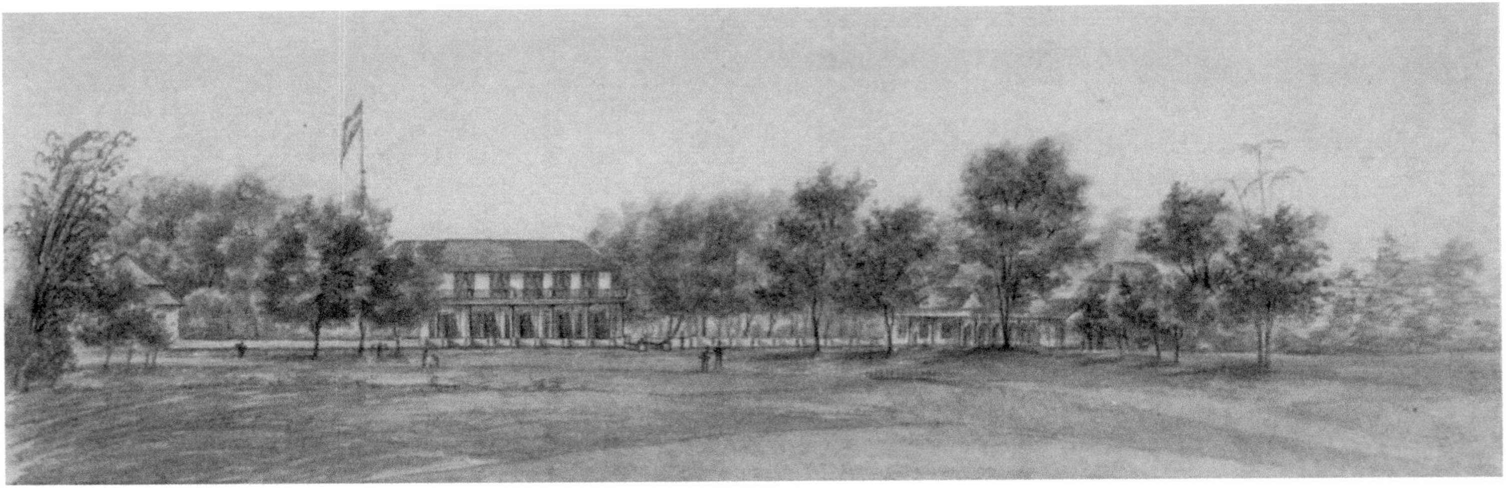

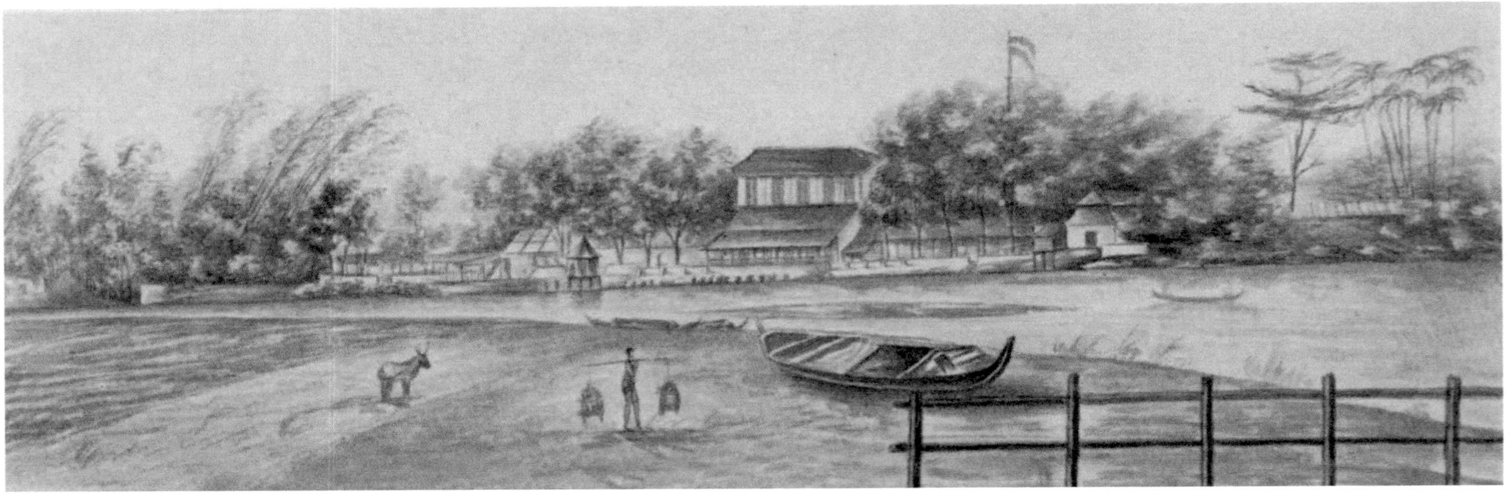

L. LE COMTE Government House, Surabaya K. 2, 19 and 20

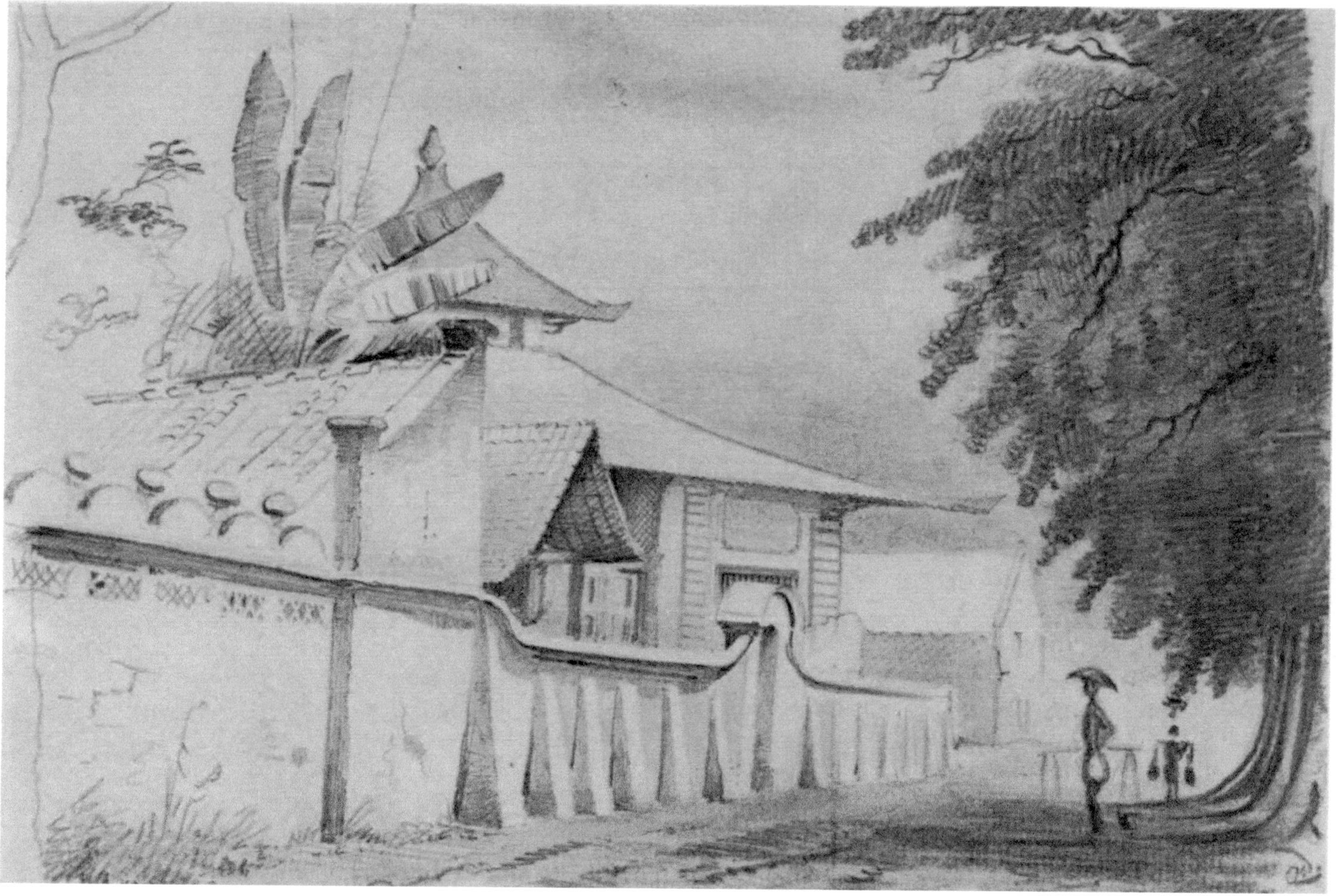

C. TH. DEELEMAN Missigit Molenvliet, Batavia K. 5, 1

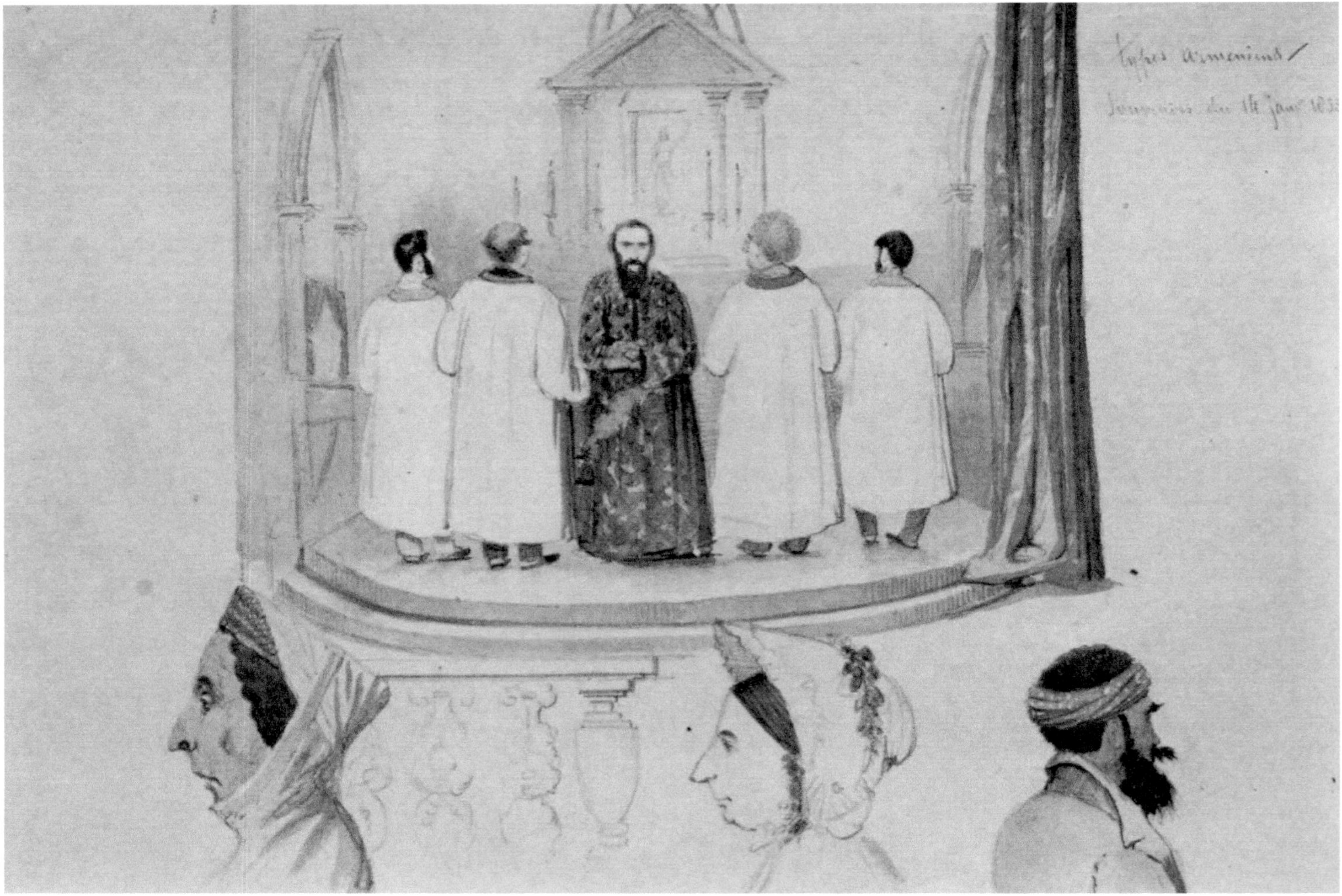

C. Th. Deeleman   Interior of the Armenian Church, Batavia   K. 5, 16

References: J. H. Maronier, Tekeningen en aquarellen van J. H. W. le Clercq, *B.K.I.* CIX (1953), pp. 227—230.

COMTE, L. le

Naval officer, one of a family of Naval officers, several members of which had a special talent for drawing. One of his relations P. le Comte (naval lieutenant 3d class 1823) published an album containing 50 lithographs of ship-models. He writes: "Als liefhebber van de kunst in dienst der Kon. Marine op mijn reizen voor mijn jeugdige beminnaars der zeevaart samengesteld" 1831 (As a lover of art serving in the Royal Navy I prepared this album during my travels for my youthful admirers of navigation, 1831). (Library of the Marine Department.)

K. 3 *Sketchbook with red binding.*
25 × 40 cm.
Note: With several detached pages K. 3, 31—42.

K. 3, 1 *Z.M. Linieschip "de Zeeuw",* Kapitn. ter zee Rijk komende op de Reede van Elzeneur en beantwoord het gedane salut der deensche Brik 1834 (H.M. Ship of the Line "de Zeeuw", Captain Rijk entering the roads of Elsinore, returns a salute from the Danish brig 1834).
Water-colours. Signed fect. L. le Comte 1835. 17,5 × 30,5 cm.
Note: In the background Kronborg Castle and the town of Elsinore.

K. 3, 2 *Z.M. Korvet Nehalennia* onder kommando van de Kapitn. Luitt. J. C. Rijk op de hoogte van Neuze de Schelden afwerkende naar Vlissingen 1832 (H.M. Corvette Nehalennia commanded by Capt. Lt. J. C. Rijk off Neuze sailing down the Schelde towards Flushing 1832).
Water-colours. Signed fec. L. le Comte 1835. 17 × 32,5 cm.

K. 3, 3 *View of coast with houses on the waterfront and churches on the hills* (Bergen?).
Rough pencil. 19,5 × 37,5 cm.

K. 3, 4 *Jogo Dollok* (Djaka-Dolok).
Water-colours. 25 × 40 cm.
Note: Djaka-Dolok is an Old-Javanese statue in Kroesenpark, Surabaya, opposite the Governer's house. The picture has been drawn on the reverse side of K. 3, 3. Concerning this statue see: N. J. Krom, *Inleiding tot de Hindoe-Javaansche kunst.* 2nd ed. 's-Gravenhage 1923, Vol. II, p. 174.

K. 3, 5 *Mosque at Palembang.*
Water-colours. 25 × 40 cm.
Note: To be compared with illustrations to the article by J. W. Wellan. "Bijdrage tot de geschiedenis van de Masdjid Lama te Palembang", *Cultureel Indië,* 1939, 305—314.

K. 3, 6 *Rivier van Macao* (River of Macao).
Water-colours. 15 × 20 cm.

K. 3, 7 *Inwoonners van Canton* (Cantonese citizens).
Water-colours. 14 × 14 cm.
Note: Nos 6 and 7 on the same sheet.

K. 3, 8 *Gezigt van Touranne Koninkrijk Siam* (View of Touranne, Kingdom of Siam (read: Indo-China)).
Water-colours. 15 × 20 cm.

K. 3, 9 *Boerinnetje van Tinneriffe* (Peasant girl of Tenerife).
Water-colours. 14 × 18 cm.

K. 3, 10 *Fisherman on the bank of a river with boats, Holland.*
Water-colours. 18,5 × 27,5 cm.

K. 3, 11 *Two cockroaches crawling and flying.*
Water-colours. 9 × 18 cm.

K. 3, 12 *Vlaardingen, quarter of the town of Macassar.*
Pencil. 4 × 20 cm.

K. 3, 13 *Vlaardingen, quarter of the town of Macassar.*
Pencil. 4 × 33 cm.

K. 3, 14 *Sketch of a Macassarese junk.*
Pencil. 14 × 23 cm.
Note: On reverse side of K. 3, 13.

K. 3, 15 *Mountain with palms and pavilion with flag (New Guinea).*
Pencil and water-colours. 18 × 40 cm.

K. 3, 16 *Mountain with palms and pavilion with flag (New Guinea).*
Water-colours. 16 × 40 cm.
Note: Same landscape as in K. 3, 15 with different details.

K. 3, 17 *View from the Hills of Giri of the town of Gresik (?) with ships in the fairway to Madura seen in background.*
Pencil. 14 × 34 cm.

K. 3, 18 *Ships off shore near Ternate.*
Water-colours. 21 × 35,5 cm.
Note: Unfinished in part.

K. 3, 19 *Rear view Government House beside Kali Mas, Surabaya.*
Water-colours. 18 × 40 cm.

K. 3, 20 *Front view of Government House, Surabaya.*
Water-colours. 18 × 40 cm.

K. 3, 21 *Bay with ships, mountains in background (Molucca's ?).*
Pencil and water-colours. 18 × 40 cm.

K. 3, 22 *View from the Hills of Giri on the town of Gresik (?) with ships in the fairway to Madura seen in background.*
Water-colours. 15 ×40 cm.
Note: Same landscape as in K. 3, 17 with more details.

K. 3, 23 *View of square with banyan tree.*
Pencil and water-colours. 12 × 40 cm.

K. 3, 24 *View of square with two banyan trees, houses and a mosque with flags.*
Water-colours. 15 ×40 cm.

K. 3, 25 *A landing stage, Macassar.*
Pencil, 10 × 40 cm.

K. 3, 26 *Pond with bathing-cabin in Java, Chinese (?) house in background.*
Water-colours. 25 × 40 cm.

K. 3, 27 *Various flowers (periwinkle (vinca rosacea), crinum kirkii* Baker, sokka (ixora Javanica), kembing merak (caesalpinia pulcherrima)).
Pencil and water-colours. 20 × 23 cm.

K. 3, 28 *Sailing-ship, three-master.*
Pencil. 10 × 15 cm.
Note: On reverse side of K. 3, 27.

K. 3, 29 *Hills near Surabaya with Chinese grave.*
Water-colours. 20 × 40 cm.

K. 3, 30 *Landfall of two islands.*
Pencil. 7 × 25 cm.
Note: With captions: 0 ½ N. Pulo Marra, O ¼ Z. Pulo Muskito.

K. 3, 31 *Het Nieuwe Diep gepeild* (Sounding the "Nieuwe Diep" (Holland)).
Pencil. Signed Le Comte 1835. 8 × 32 cm.

K. 3, 32 *Naval Headquarters, den Helder, with mast of gunboat No. 4 van Speyk.*
Water-colours. 25 × 37 cm.

K. 3, 33 *The town of Medemblik with Neth. Reformed church.*
Pencil. 15 × 32 cm.

K. 3, 34 a. *Beacon.* Pencil. 17 × 10 cm.
b. *Gate.* Pencil. 15 × 12 cm.
c. *Fishing-boats.* Pencil. 3,5 × 7 cm.
d. *Royal Marine Institute.* Pencil. 1 × 7 cm.

K. 3, 35 *View of the town of Medemblik from the West.*
Pencil, grey wash. 18 × 34 cm.

K. 3, 36 *Urk van den vuurtoren gezien* (View of (the isle of) Urk from lighthouse).
Pencil. 17 × 34 cm.
Note: On reverse side of K. 3, 35.

K. 3, 37 *Castle of Radboud, before its restoration, Medemblik.*
Pencil. 17 × 29 cm.

K. 3, 38 *View of the town of Medemblik from the East.*
Pencil. 12 × 20 cm.
Note: On reverse side of K. 3, 37.

K. 3, 39 *Hull of sailing-ship, with figure-head of 17th century warrior.*
Water-colours. 12 × 34 cm.

K. 3, 40 *Hull of sailing-ship, with figure-head of old-fashioned soldier with shield and rapier.*
Water-colours. 13 × 35 cm.

K. 3, 41 a. *Woman's head.* Pencil. 6 × 5 cm.
b. *Naval officer.* Pencil. 12 × 12 cm.
Note: With captions: Pulo Muskito klein en laag (Pulo Muskito small and low), Pulo Tiya laag en boomrijk met een rif ten Zuiden op 1¼ mijl afstand (Pulo Tiya low and well-timbered with a reef at a distance of 1¼ mile to the South), Pulo Douma laag en boomrijk (Pulo Douma low and well-timbered), Pulo Sinnaro klein laag en boomrijk (Pulo Sinnaro small low and well-timbered).

K. 3, 42 *Boomjes Eiland* (Boomjes (= tree) island (landfall)).
Pencil sketch. 6 × 25 cm.
Note: On reverse side of K. 3, 41.

Provenance: Bought from B. Harteveld, Leiden.

DEELEMAN, Charles Théodore

Born Dec. 11,1823, Amsterdam. Died Febr. 13, 1884, Batavia. Studied engineering in Delft. Went to Indonesia in 1845 as an engineer with the Neth.-Ind. waterways service. In Surabaya until 1848, after that in Batavia. In 1853 he designed the building for the exhibition of Batavia. Joined the firm of Chaulan and Deeleman as a partner in 1854. Designer of a horse-drawn carriage named "deeleman" after him. Made many water-colours and pencil-drawings of the Indonesian landscape. He was, moreover, not without merit as a caricaturist. Coloured lithographs of his sketches were printed in *Bataviaasch Album, verzameling van een tiental gezichten van de hoofdstad van Nederlandsch Indië* (Batavia Album, collection of ten views of the capital of the Neth. Indies), Batavia [1859]. He contributed also to the collection of prints published under the title *De Indische Archipel* (The Indonesian Archipelago), 's-Gravenhage 1877.

K. 5, 1 *Missigit Molenvliet, Batavia.*
Pencil. Signed CD. 16,5 × 19,5 cm.
Note: On reverse side 2 caricatures, pencil.

K. 5, 2 *Waterloo Square with decorated Waterloo Memorial, Batavia.*
Pencil. 16,5 × 25,5 cm.
Note: On reverse side 5 caricatures, one of which is of Dr. E. R. Kielstra, pencil.

K. 5, 3 *Bridge Goenoeng Sahari, Batavia.*
Pencil. 16,5 × 22,5 cm.

K. 5, 4 *Salatiga.*
Pencil. 16,5 × 24,5 cm.
Note: View of Salatiga with the Dutch fort "de Hersteller".

K. 5, 5 *Road from Bandung to Sumedang.*
Pencil. 16,5 × 25 cm.
Note: Two drawings, marked: Paal 134. Paal 135 (milestone 134, ditto 135).

K. 5, 6 *Military school at Meester Cornelis.*
Pencil. 16,5 × 28 cm.

K. 5, 7 *Caricature of L. W. C. Keuchenius.*
Water-colours on front and reverse sides. 17,5 × 6 cm.
Note: Marked on one side: Bataviaasch H[andels]blad en Nieuw Bat. Ha[ndelsb]lad "Laïn sekarang Laïn doolo" (Batavian Commercial newspaper and New Bat. Commercial newspaper "Today is unlike yesterday".) Marked on the other side: Druk[persre]glement, gewro[ken erg]ernis (?) "Laïn doolo Laïn sekarang". (Press regulations, irritation revenged "Yesterday was unlike today".) With letter from Keuchenius to Deeleman. Aug. 24, 1870, Batavia.

K. 5, 8 *Four drawings illustrating the use of various two-wheeled carriages.*
Ink and pencil. All 13 × 18,5 cm.
Note: With captions:
Fig. 1 "Geoorloofd en voorgeschreven maar geschied nooit." (Permitted and prescribed but never done.) On the inside: A "Strikt genomen geoorloofd; over de beenen wordt niets gezegd. Welk een magt over het paard." (Strictly speaking permitted; nothing is said about the legs. What control of the horse!) B "Ongeoorloofd en geschied altijd." (Not permitted and always done.)
Fig. 2 "Ongeoorloofd en geschied veelal." (Not permitted and mostly done.) On the inside: "Geoorloofd en gebeurd nooit". (Permitted and never done).
Fig. 3 On top: "Geoorloofd en élégant". (Permitted and elegant.) Underneath: "Niet gevaarlijk, en met een groote magt over (de staart van) het paard". (Not dangerous and with great control of the horse ('s tail).)
Fig. 4 On top: "Ongeoorloofd . . (.) maar, dit is noch bendy noch dos-à-dos". (Not permitted . . (.) but, this is neither a bandy nor a *dos-à-dos).* Underneath: " "Batavienne" Handwagentje op 2 wielen". (Two-wheeled Batavian buggy.) "Deze teekening is oorzaak dat de bepaling (.) dat geene voertuigen van achteren mogen gereden worden, is ingetrokken". (Because of this drawing the rule against driving a vehicle from the rear has been cancelled.)
With letter from the resident of Batavia to Deeleman. April 17, 1863, Batavia.

K. 5, 9 *Edele wedstrijd voor het Kommandement van het Helsche Peloton van 14 rotten* (Honourable contest for the command of the Hellish Platoon of 14 squads).
Pencil. 26,5 × 19,5 cm.
Note: Perhaps Gov. Gen. P. Mijer and the High Council of the Indies?

K. 5, 10 *Ootmoedig verzoek aan Uwe Exce. om te worden bevorderd tot Postpaard* (Humble request to Your Excellency to be promoted to be posthorse).
Ink. 10 × 16,5 cm.
K. 5, 11 *Ontwerp Monument ter herinnering aan het overlijden der Maatschappij van "Toonkunst" op 7 Maart 1877* (Design for Monument to commemorate the death of the Musical Society on 7th March, 1877).
Ink. 32 × 21,5 cm.
Note: Uppercorners cut off obliquely. On pedestal of the monument: T. F. Marcus Luitt. Kapelmeester der Stafmuzijk. Off. der Orde van het Grootste Witte Beest. Directeur der Bataviasche Muzijk en Zang Vereeniging enz. enz. enz. (T. F. Marcus Ltn. Bandmaster of Regimental band. Off. of the Order of the Greatest White Beast. Director of the Batavian Singing and Music club etc. etc. etc.) With marginal note: Zie Javabode 7 Maart 1877 (See *Javabode* 7th March, 1877).

K. 5, 12 *Cavalcade.*
Pencil. 16,5 × 24 cm.
Note: With caricature of Deeleman.

K. 5, 13 *Caricature of Whist party.*
Pen and sepia. 23 × 29 cm.
Note: The captions are Dutch puns, impossible to translate. Probably has some connection with K. 5, 9. Is this the High Council of the Indies?

K. 5, 14 *Caricatures, among which a musician.*
Water-colours and pencil. 24 × 18,5 cm.

K. 5, 15 *Pasanggrahan* (Rest-house).
Water-colours. Signed Deeleman. Ca. 1860. 16 × 27,5 cm.

K. 5, 16 *Interior of the Armenian Church in Batavia.*
Water-colours and pencil. 16,5 × 24 cm.
Note: With marginal note: Types armeniens. Souvenirs du 14 Janv. 1855. (Armenian types, memento of 14th Jan., 1855.)
H. 654 a. A French poem by Deeleman:
"Complainte sur un horrible crime commis à Java et dont l'accusée, condamné à être pendue il y a . . . mois, se porte aujourd'hui comme vous et moi, composée et illustrée par le Baron Caperdu-Laboula". Illustrated by him with 23 caricatures.
b. A printed occasional poem on the marriage of Jan van Drooge also contains 9 caricatures by Deeleman.
c. A French poem by Deeleman "Le Paradis" with one drawing.

Provenance: All water-colours and drawings, except K. 5, 15, were presented to the Institute by Mr. A. W. Deeleman by courtesy of

Ir. L. V. Joekes. K. 5, 15 was bought at Waterloo Square market, Amsterdam, by Dr. H. J. de Graaf.

DEKENS, Adolphe

Born 1820, Gent. Went to the Neth. Indies as a painter Dec. 5, 1840. Left Surabaja and settled in Semarang Nov. 1, 1842.

S. 4 *Ruin of Tjandi Mendoet before the restoration.*
Water-colours. Signed A. Dekens 1850. 37 × 44,5 cm. In gilt frame 54 × 66 cm.
Note: Probably the first, in any case one of the earliest pictures of this Buddhist temple, as its existence was made public only in the year 1838. (N. J. Krom, *Inleiding tot de Hindoe-Jav. kunst.* Vol. I, p. 295.)

ERP, Dr. Th. van

Born March 26, 1874, Amboina. Died May 7, 1958, Laren.
As a Lieut. engineer of the K.N.I.L. (Royal Dutch Indies Army) he was appointed as a member of the Borobudur-Committee. After he had handed in his report in 1907 he was commissioned to supervise the restoration of the Borobudur. After its completion in 1911 he joined Prof. Dr. N. J. Krom in preparing the publications of the monumental work *De archeologische beschrijving van de Baraboedoer* (The archeological description of the Borobudur). This contains many drawings and ca. 1460 photographs made by Van Erp or under his supervision. He also restored other temples in Central Java. Unfortunately many of his drawings and paintings were destroyed during a bombing raid on The Hague, 1945. See *Indonesië,* Vol. 5, 1951, pp. 279—284.

S. 20 *Design for block representing the emblem of the Institute (çankha-motive) used on all its publications.*
Ink. 1938. 22 × 22 cm.

F. 4, 1 *Borobudur, upper galleries with procession.*
Chalk on tinted paper. Pasted on cardboard. 24 × 32 cm.

F. 4, 2 *Tjandi Mendoet, interior view with worshippers.*
Chalk on tinted paper. Pasted on cardboard. 24 × 32 cm.
Note: Drawn freely from nature.

Provenance: Presented by the artist.

FAHRINGER, Carl

Born Dec. 25, 1874, Wiener Neustadt. Died Feb. 4, 1952. Vienna. Was trained at the Allgem. Malerschule by Prof. Chr. Griepenkerl and at the Spezial Schule by Prof. Aug. Eisenmenger in Vienna and later in München. Made study-tours of Italy, France, Egypt, Africa, Java, Bali, Sumatra and Greece. Was appointed professor at the Allgem. Malerschule der Akademie der Bildende Künste, Nov. 1, 1929.

S. 5 *Meru temple with shrines in foreground. Bali.*
Oil. Signed C. Fahringer. 64 × 44 cm. White frame.

S. 6 *Kingsgraves Tampah Siring. Gjanjar. Bali.*
Oil. Signed C. Fahringer. 23 × 33 cm. Black frame.

S. 7 *Temple entrance with shrines. Bali.*
Oil. Signed C. Fahringer. 36 × 25 cm. Black frame.
Note: In right lower corner KAWI.

K. 6, 1 *Squatting Balinese boy.*
Chalk. Dated 'Bali '27'. 22 × 18 cm.

Provenance: Presented by Mrs. I. F. Damsté-Muller.

FLEISCHER, Max

Born July 4, 1861, Lipin (Silesia). Died April 3, 1930, Menton. Trained in Breslau, Berlin and München. Visited Indonesia frequently, viz. 1899, 1903, 1909, and 1931—1918. Exhibited his work in the Netherlands Indies Art Society, 4th—17th July, 1919. Many of his pictures are in possession of the Royal Tropical Intitute (Amsterdam). He published a monograph on Javanese mosses.

GF. 1, 3 *Barabudur.*
Water-colours. Signed Max Fleischer 1913. 30 × 45,5 cm.
Note: One of the upper terraces.

Provenance: Presented by J. H. Maronier. 1959.
References: Weekblad voor Indië 1909—'10, p. 257—261.

GAFFRON, Heinrich von

Born April 28, 1813, Grebel(witz?). Died Nov. 2, 1880, Sindanglaja. Studied at the Military Academy of Berlin, was 2d lieutenant in the Prussian Army. Went to Neth. Ind. in 1839 as a soldier, later on he was appointed gunner. In 1840 he joined, as a draughtsman, Dr. E. A. Forsten, member of the Commission for Scientific Research. They travelled through the Minahassa and the Moluccas. After the death of Dr. Forsten Jan. 3, 1843, Von Gaffron joined another member of the same commission, Dr. C. A. L. M. Schwaner, and went with him on several expeditions, prospecting for coal. He also travelled on his own. On May 19, 1846 he was put at the disposal of the Governor of Borneo, A. L. Weddik, and was charged with the preparations for a new settlement at Kota Waringin. In that function he travelled through South-West Borneo and probably Serawak. In 1848 he discovered the coal-layers of Pengaron and was appointed administrator of the new coal-mine Oranje Nassau in 1849. In 1851 he qualified as a minor official and from then on made his carreer in the civil service. In 1856 he was acting resident of Sintang, probably till 1860, and distinguished himself in a battle with Malays and Dayaks. These exploits were rewarded with a letter of commendation and a military cross, the Willemsorde 4e klasse.

1860—1868 assistant resident of Biliton. Member of the Society of Scientific Research. May 31, 1858, the Minister of the Colonies entrusted the drawings of Von Graffon and those of Schwaner and Aug. Pers to the Royal Inst. on the proposal of the Governor-General at Buitenzorg, because these 108 drawings, still in possession of the Inst., might be needed for a new edition of C. A. L. M. Schwaner's book *Borneo, beschrijving van het stroomgebied van den Barito en reizen langs eenige voorname rivieren van het Zuid-Oostelijk gedeelte van dat eiland . . . op last van het Gouvernement van Nederl. Indië gedaan in de Jaren 1843—1847 met platen en eene kaart.* Uitgegeven vanwege het Kon. Instituut . . . te Amsterdam . . . 1853—'54. 2 dln. Some of the lithographs in this book are made after sketches of H. von Gaffron. His drawings are catalogued as systematically and chronologically as possible.

Dayak types from the Kapuas Murung River area.

K. 7, 1 *Dayak in war array.*
Pencil. 28 × 19 cm.
Note: See picture "Dajak der Doeson-landen in oorlogs-tenue" in *Ethnographische beschrijving der Dajaks* (p. 62) door M. T. H. Perelaer ("Dayak from the Duson districts in war array" in Ethnographical description of the Dayaks (p. 62) by M. T. H. Perelaer).

K. 7, 2 *Daijaksche Frauen von Poelo Pedak* (Dayak women of Pulo Pedak).
Pencil. Signed H. v. Gaff. 25 × 18 cm.

K. 7, 3 *Eine alte Daijakker Frauw von Kotta Baroe* (An old Dayak woman of Kota Baru).
Pencil. 29 × 16,5 cm.

K. 7, 4 *Eine Daijaksche Frau von der Arbeit kommend. Kotta baroe* (A Dayak woman returning home from work. Kota Baru).
Pencil. 22,5 × 17 cm.

K. 7, 5 *Kleidung der Daijakschen Frauen im Hause am Sg. Kapoas* Dayak women's clothing for wearing at home, on the Kapuas River).
Pencil. Signed H. v. G. 26 × 15 cm .

K. 7, 6 *Daijaksche Frauen von Karangan am Kapoas* (Dayak women of Karangan on the Kapuas).
Pencil. Signed H. v. Gaffron 1845. 29 × 19 cm.

K. 7, 7 *Ein gewafneter Daijakker von Kotta baroe. Kapoas fl* (Armed Dayak of Kota Baru, Kapuas River).
Pencil. 27,5 × 19 cm.

K. 7, 8 *Ein Daijakker von Kotta baroe am Kapoas* (Dayak of Kota Baru on the Kapuas).
Pencil. 24,5 × 18 cm.

K. 7, 9 *Daijaker Paree* (Dayak men of Paree).
Pencil. 19,5 × 29,5 cm.

K. 7, 10 *Szene aus ein Daijjakker feste; im Vordergrunde ein Bassier* (Scene from a Dayak feast with a 'basir' (priest) in foreground).
Pencil. 26 × 22 cm.

K. 7, 11 *Bakadza.*
Pencil. 17 × 14 cm.

K. 7, 12 *Dayak with shield and spear.*
Pencil. 25 × 15,5 cm.

K. 7, 13 *Details of men's dress from the Kapuas Murung.*
Pencil. 22 × 14 cm.

Drawings of Banjarmasin and environs.

F. 5, 1 *Banjermassing.*
Water-colours. Signed H. v. Gaffron. 24,5 × 35 cm.
Note: In lower left corner co-signature: Banjermass. 6/12 '43 Dr. C. M. Schwaner.

F. 5, 2 *Südlicher Theil von Banjermassing 1843* (Southern part of Banjarmasin 1843).
Water-colours. Signed H. v. Gaffron. 24,5 × 35 cm.

F. 5, 3 *Schanze von Thijl am Flusse von Banjermassing* (Fort van Thijl (read: Tuijl) on the river of Bajarmasin).
Water-colours. Signed H. v. Gaffron. 25 × 34,5 cm.
Note: Co-signature: Banjermass. 25/11 '43 Dr. C. M. Schwaner.

K. 7, 14 *Triumphal chariot.*
Water-colours. Signed H. v. Gaffron. 20 × 31 cm.
Note: Probably the sultan of Banjarmasin with his wife and the Governor, depicted sitting in this six-wheeled vehicle which is drawn by his subjects.

K. 7, 15 *Box in the shape of a Garuda.*
Ink. Signed H. v. Gaffron Banjermassing 1843. 21,5 × 34,5 cm.
Note: Co-signature: Banjermass. 6/11 43 Dr. C. M. Schwaner.

K. 7, 16 *Reiss-Floss, Raquet genannt wie es im innern des Reiches von Banjermassing üblich sind* (Travelling raft, called raquet, as used in the interior of the Banjarmasin state).
Pencil. 17 × 25 cm.
Note: According to this drawing, a "Reiss-Floss" seems to be a travelling-raft (Mal. *rakit*) generally used for travelling in the interior of the sultanate of Banjar Masin, in this case by a European (official?) whose legs are seen projecting from the side-door.

K. 7, 17 *Kampong with guard at the Dutch flag.*
Pencil. Signed Gaffron. 19 × 27 cm.

Tabanio.

K. 7, 18 *Tabanio, fort and civil service post* (No. 1a.).
Pencil. 20,5 × 31,5 cm.

K. 7, 19 *Tabanio, grounds to the South of the civil service post* (No. 1b.).
Pencil. 18 × 31,5 cm.
Note: These drawings, 1a and 1b, fit together.

F. 5, 4 *Tabanio, Borneo.*
Pencil. Signed H. v. Gaffron. 20 × 35 cm.
Note: Much the same as K. 7, 18.

F. 5, 5 *Tabanio, seen from the river.*
Pencil. Signed H. v. Gaffron Tabanio 1844. 22,5 × 36,5 cm.

Drawings of scenes from geological prospecting expeditions in South-East Borneo, about 1844—'45.

K. 7, 20 *Ite Schichte* (First coal-layer).
Pencil. 12,5 × 20,5 cm.
Note: Coal-layer near the Gunung Batu Api.

K. 7, 21 *2te Schichte* (Second coal-layer).
Pencil. Signed Gaffron 16/1, 44. 13,5 × 20,5 cm.
Note: Coal-layer near the Gunung Batu Api.

K. 7, 22 *IIIde Kolenlaag aan de goenong Batoe Apie* (Third coal-layer near the Gunung Batu Api).
Ink and pencil. Signed H. v. Gaffron 4/3, 44. 20,5 × 30,5 cm.

K. 7, 23 *Bog-ore at Sungei Basar.*
Pencil. Signed H. v. Gaffron Soengij Basar d. 11/7, 44. 25 × 19,5 cm.

K. 7, 24 *Atanik aan de Sungij Batoe Apie* (Atanik on the Sungei Batu Api).
Pencil. Signed Gaffron 1844. 21,5 × 32 cm.

K. 7, 25 *Goe. Seniang* (Gunung Seniang).
Pencil. Signed H. v. Gaffron. 18,5 × 25 cm.
Note: The prospectors Schwaner and Gaffron are shown surveying the land.

K. 7, 26 *Ansicht der Goe. Ratus südl. von Kamp. Tombaga* (View of the Gunung Ratus to the South of Kampong Tombaga).
Pencil. Signed v. Gaffron. 19,5 × 25 cm.
Note: Captions added later on: Berggezigten in de Tanah Laut I (Mountain views in Tanah Laut I). Gezigt op het Hondertgebergte bezuiden de Kampong Tambage (View of the Hundred Mountains South of Kampong Tambaga). Besides other notes.

K. 7, 27 *Gezigt van de Kust der Tanah laut genomen van Poelo Dato* (View of the shore of Tanah Laut from Pulo Dato).
Pencil. 11 × 19,5 cm.

K. 7, 28 *Gezigt op de Goe. Ratus van uit Poelo Sarij (Lampeij)* (View of the Gunung Ratus from Pulo Sari (Lampey)).
Pencil. Signed H. v. Gaffron 1844. 15 × 24 cm.

K. 7, 29 *Steenkolenschichten aan de Sg. Lampang* (Coal-layers on the Sungei Lampang).
Ink. 9,5 × 17 cm.

K. 7, 30 *Profil van de Steenkolen in het streiken van dezelven tegen Z.W.t.Z.* (Cross-section of coal-layer to the S.W.S.).
Ink. 11,5 × 17 cm.
Note: Cross-section of K. 7, 29. In lower right corner: "Opgenomen en geteekend door H. v. Gaffron den 27e Januarij 1845" (surveyed and drawn by H. v. Gaffron 27 Jan., 1845).

K. 7, 31 *Kamp. Baroe bij Pagatan* (Kampong Baru near Pagatan).
Pencil. Signed H. v. Gaffron. 18,5 × 27,5 cm.

K. 7, 32 *Het vorstelijke paleis van Pagattan (Boeginezen)* (The royal palace at Pagatan (Bugis)).
Pencil. 19,5 × 24 cm.

K. 7, 33 *Glück auf!* (Good day!).
Water-colours. Signed H. v. Gaffron. 19,5 × 27,5 cm.
Note: The drawing shows European miners near their mine beside the river.

Note: The numbers K. 7, 26, 27, 28 have been copied on the map of C. M. Schwaner's Topografische en Statistische Reizen (Topographical and Statistical expeditions), p. VIII (MS. in Leiden Univerity Library). Dutch titles have been added under the common name: Gezigten in de Tanah Laut (Views of the Tanah Laut).

Journey through the Kapuas Murung River basin, about 1845.

K. 7, 34 *Catching fish on the Kapuas Murung River.*
Pencil. 17,5 × 22 cm.

K. 7, 35 *Landscape, Kapuas Murung River basin.*
Pencil. 18 × 29 cm.

K. 7, 36 *Landscape, Kapuas Murung River basin.*
Pencil. 20,5 × 31 cm.

K. 7, 37 *Sketch of a house, Kapuas Murung River basin.*
Pencil. 16 × 30 cm.

K. 7, 38 *Kotta Tapen am Kapoas* (Kota Tapen on the Kapuas Murung River).
Pencil. 15 × 22 cm.

K. 7, 39 *Tapen am Kapoas* (Tapen on the Kapuas Murung River).
Ink. Dated 4/6, 45. 12 × 18 cm.
Note: Same subject as K. 7, 38.

K. 7, 40 *Benteng Koharingan (Karangan?) on the Kapuas.*
Pencil. Signed H. v. G. 18,5 × 31,5 cm.

F. 5, 6 *Innere Ansicht der Benteng Koharingan* (View of the interior of the benteng Koharingan (Karangan?)).
Pencil. 21,5 × 35 cm.
Note: Reproduced as a lithograph in *Neerlandsch Oost-Indië* by S. A. Buddingh, Vol. III, facing p. 421, with caption: Kotta Karingan.

K. 7, 41 *Das Innere von einer Daijakschen Benteng* (Interior of a Dayak benteng).
Pencil. 19,5 × 26 cm.
Note: Reproduced as a lithograph in *Borneo* by Schwaner, Vol. II, facing p. 22, with caption: Een Daijaksch Dorp (Benting) (a Dayak village (benteng)).

K. 7, 42 *Tanz der Daijakker am Kapoas* (Dayak dance in the Kapuas River area).
Pencil. 19,5 × 29,5 cm.
Note: Reproduced as a lithograph in *Borneo* by Schwaner, Vol. II, facing p. 150, with caption: Een dans der Ot: Danomsche mannen (A dance of Ot Danum men).

K. 7, 43 *Tanz Daijakker Frauen* (Dance of Dayak women).
Pencil. 21,5 × 31,5 cm.

K. 7, 44 *Verbrennen des Leichen bei den Daijakkern zu Koharingan* (Dayaks of Koharingan (Karangan?) cremating corpses).
Pencil. 21,5 × 32 cm.
Note: Reproduced as a lithograph in *Borneo* by Schwaner, Vol. II, facing p. 76, with caption: Het verbranden der lijken aan de Kapoeas Moeroeng (Cremating of corpses on the Kapuas Murung).

K. 7, 45 *Burial-platforms in the Kapuas Murung River area.*
Pencil. 14 × 29 cm.

F. 5, 7 *Daijaksche Gräber am Kapoas* (Dayak burial platforms in the Kapuas (Murung) River area).
Pencil. 19 × 35 cm.

Journeys through the Dusun and (Siang) Murung districts, 1845 and 1846(?).

F. 5, 8 *Bliang-scene am Pattaij* (Scene from a Bliang festival on the Patai).
Pencil. 22 × 36,5 cm.
Note: Reproduced as a lithograph in *Borneo* by Schwaner, Vol. II, facing p. 44, with caption: Een bilian feest der Daijakkers van Soengei Pattaij (A Dayak balian feast of the Sungei Patai area). The bilians are described in Vol. I, p. 185—186. Von Gaffron stayed in Patai from 19th—21st July, 1845. The Patai is a tributary of the Barito River.

F. 5, 9 *Innere eines Daijakker hauses am Karrau* (Interior of a Dayak house in the Karau River area).
Pencil. 22,5 × 36,5 cm.

Note: Reproduced as a lithograph in *Borneo* by Schwaner, Vol. II, facing p. 182, with caption: Een Daijaksche woning van binnen (Interior of a Dayak house). Von Gaffron visited Karau on July 27, 1845. The Karau is a tributary of the Barito River.

K. 7, 46 *Benting Rapen, gegenüber Latoentoer am Baritoe* (Fort Rapen opposite Latoentoer (Loluntuor?) on the Barito).
Pencil. Signed H. v. Gaffron 1846. 21 × 32 cm.
Note: Von Gaffron visited Loluntuor on Feb. 4—5, 1846.

K. 7, 47 *Eisenschmelze in der Doeson* (Iron foundry in Dusson (district Ulu)).
Pencil. Signed H. v. Gaffron. 20,5 × 28,5 cm.
Note: Reproduced as a lithograph in *Borneo* by Schwaner, Vol. I, facing p. 110, with caption: Eene Daijaksche ijzersmelterij aan de Barito (A Dayak iron foundry on the Barito). Iron-founding is described in Vol. I, p. 109—113. Von Gaffron visited an iron foundry at the Kampong Bodje on Feb. 3, 1846, and described it in his diary.

K. 7, 48 *Bahan in Moerong am Sg. Baritoe* (Bahan in (Siang) Murung on the Barito River).
Pencil. Signed H. v. Gaffron. 18 × 28 cm.
Note: Von Gaffron visited Bahan on Feb. 8—11, 1846.

F. 5,10 *Benteng Jager am Baritoe zu über Bahain in Moerong* (Benteng Jager on the Barito River opposite Bahan in (Siang) Murong).
Pencil. 22 × 48,5 cm.
Note: Von Gaffron visited Bahan on Feb. 8—11,1846.

Animals.

K. 7, 49 *Sheet with studies of animals.*
Ink and water-colours. Signed H. v. Gaffron Banjermassing 1843. 24 × 17,5 cm.
Note: Fam. Helicoideae (actual size), Gasteropoda (actual size), Arachnidae Fam. Araneae.

K. 7, 50 [*Bekerzwam*].
Water-colours. Signed H. v. Gaffron 11/12 43. 18 × 20 cm.

K. 7, 51 *Garfish* (actual size).
Ink and pencil. Signed H. v. Gaffron Banjermassing 44. 20 × 22,5 cm.

K. 7, 52 *Pipe-fish* (actual size).
Ink and pencil. Signed H. v. Gaffron Banjermassing 44. 20,5 × 26,5 cm.

K. 7, 53 *Catfish* (1/3 actual size).
Pencil. Signed H. v. Gaffron Banjermassing 1844. 20 × 31 cm.

K. 7, 54 *Serpentine animal.*
Water-colours. Signed H. v. Gaffron Banjermassing 4/1, 44.
22 × 25,5 cm.

K. 7, 55 *Draco volans.*
Water-colours. Signed H. v. Gaffron 1844 Martapura. 24,5 × 17 cm.

F. 5, 11 *Orang-Utans near the river.*
Water-colours. Signed H. v. Gaffron 19/4, 44. 25 × 33,5 cm.

F. 5, 12 *Orang-Utan.*
Pencil. Signed H. v. Gaffron Banjermassing 1844. 25 × 35,5 cm.

Plants. Drawings probably made on the 2nd journey to the Duson and (Kapuas) Murung districts, in the early part of 1846.

K. 7, 57 *Kajoe Zamara No. 1.*
Pencil. 15 × 20 cm.
Note: With captions in German, German script.

K. 7, 58 *Kajoe Sampan No. 2.*
Pencil. 15 × 20 cm.
Note: With captions in German, German script.

K. 7, 59 *Kajoe jamplong No. 3. Kajoe Baroe laut No. 4.*
Pencil. 15 × 20 cm.
Note: With captions in German, German script.

K. 7, 56 *Kajoe Bintara of Klampang No. 5* (Kayu Bintara or Klampang No. 5).
Pencil. 15 × 20 cm.

K. 7, 60 *Kajoe Bankal No. 10.*
Pencil. 22 × 28 cm.

K. 7, 61 *Kajoe Awating 11.*
Pencil. 18 × 27,5 cm.

K. 7, 62 *Kajoe Mampeij 12.*
Pencil. 18 × 27,5 cm.

K. 7, 63 *Kajoe Gallam of kajoe poetie No. 14* (Kayu Gallam or kayu putih).
Pencil. 19,5 × 25,5 cm.

K. 7, 64 *Plandukan oder Tallie houtang* (Plandukan or Tali hutan).
Pencil. 15 × 20 cm.
Note: With captions in German, German script.

K. 7, 65 *Kajoe Baru.*
Pencil. 15 × 20 cm.

K. 7, 66 *Branch with leaves and fruit.*
Pencil. 19,5 × 23,5 cm.
Note: With caption in German, German script.

F. 5, 13 *Studies of various plants of Borneo.*
Pencil. 22,5 × 34 cm.
Note: With captions.

Serawak, about 1847.

K. 7, 67 *Serawak.*
Pencil. Signed H. v. Gaffron. 17 × 22,5 cm.
Note: The bungalow of Mr. J. Brooke is in the background.

K. 7, 68 *Serawak.*
Pencil. Signed H. v. Gaffron. 15,5 × 26 cm.

K. 7, 69 *Das Innere von einem Hause zu Randobokula am Fusse des Kristalgebirges* (Interior of a house in Randobokula at the feet of the Crystal Mountains).
Pencil. Signed Gaffron. 18 × 22,5 cm.

F. 5, 14 *Jankan.*
Pencil. 22 × 34,5 cm.

F. 5, 15 *Kampement aan de Sg. Bassar* (Camp beside the Basar River).
Ink, pencil and water-colours. Signed H. v. Gaffron. 25 × 37 cm.

F. 5, 16 *Waterval aan de Sungij Pinang* (Waterfall in the river Pinang).
Pencil. Signed Gaffron 1844. 24,5 × 34 cm.

F. 5, 17 *Kiaij 1844.*
Water-colours. Signed H. v. Gaffron 1844. 24,5 × 34 cm.
Note: The drawing shows the reception given the members of the expedition.

F. 5, 18 *Gezigten van het Zuid-Kaap en een gedeelte der West-Kust der Tanah Laut* (Views of the South-Cape and part of the West-coast of Tanah Laut).
Ink. Signed H. v. Gaffron Banjermassing 1843. 29 × 35,5 cm.
Note: 4 drawings with captions. Co-signature: Banjermassing am 26 Octob. 1843 C. M. Schwaner.

F. 5, 19 I. *Gezigt op het Gebergte tegen Z.W. tot N.W. van de Goenoeng Batoe Apie.* II. *Gezigt op het Gebergte van het N.W. tot tegen O. en O.S.O. van de goenoeng Batoe Apie* (Views of the mountains S.W. to E.S.E. of the Gunung Batu Api).
Ink. Signed H. v. Gaffron Banjermassing 1844. 23,5 × 68 cm.

GALESTIN, Dr. Th. P.

Born June 5, 1907.
He took his doctor's degree at the University of Leiden in 1936. Was appointed Professor at Amsterdam University, went to Leiden in 1958 as Professor of archaeology and history of S. and S.E. Asia. Made several study-tours to Bali where he made excellent portraits of various native types. The illustrative drawings made by him for inclusion in *Java in the 14th century* by Th. G. Th. Pigeaud were granted to the collection of the Institute.

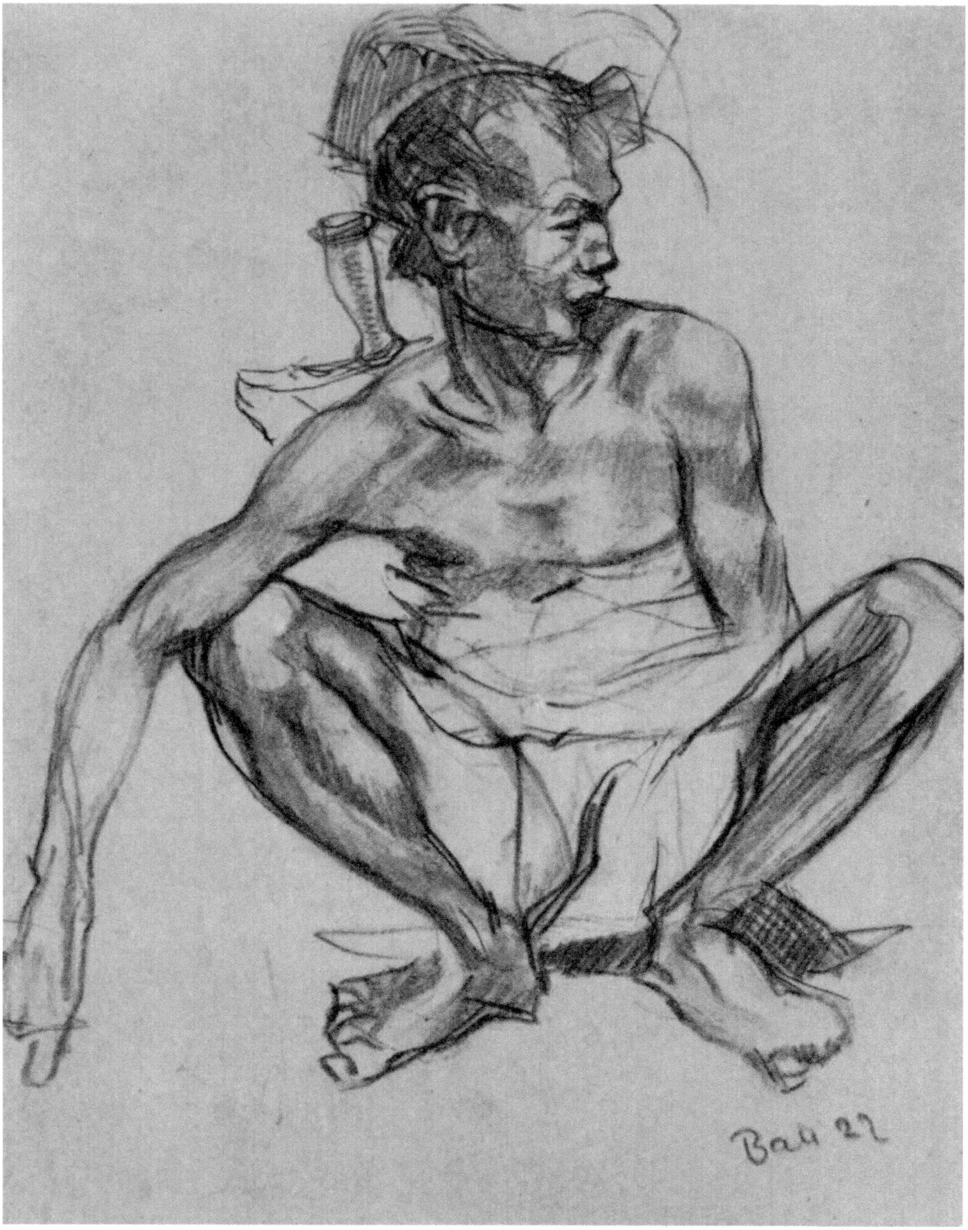

C. Fahringer Squatting Balinese boy K. 6, 1

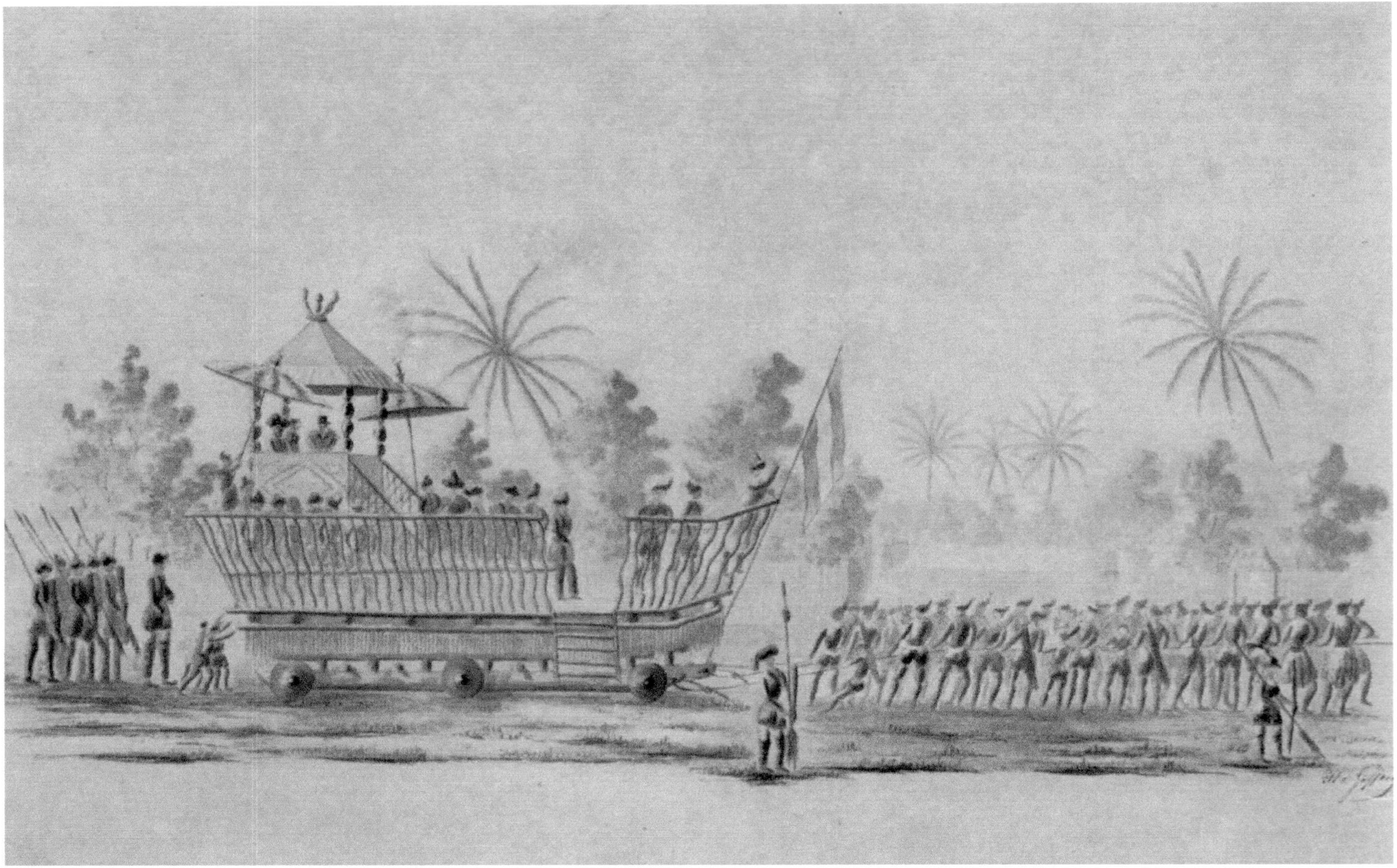

H. von Gaffron Triumphal chariot K. 7, 14

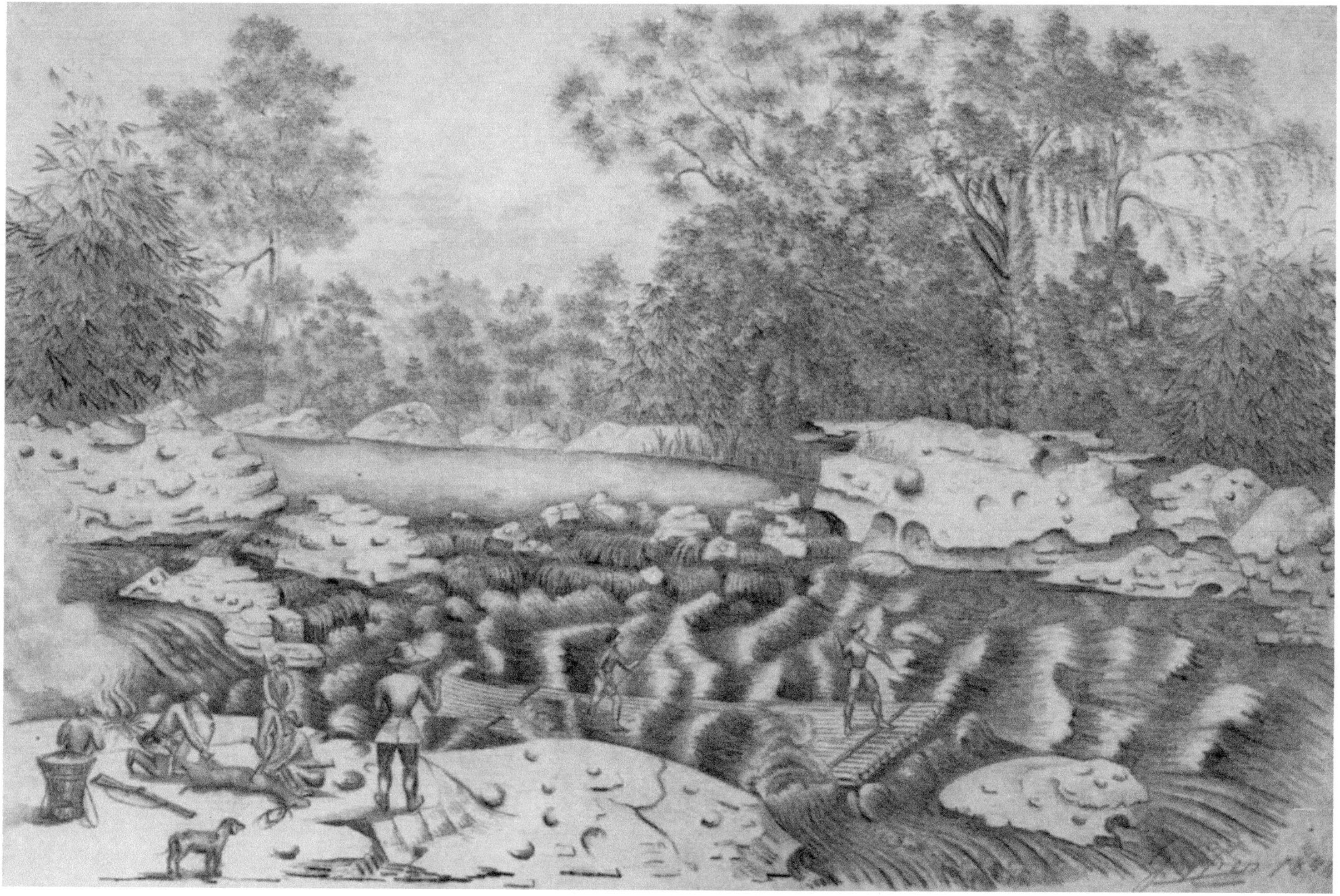

H. von Gaffron   Waterfall in the river Pinang   F. 5, 16

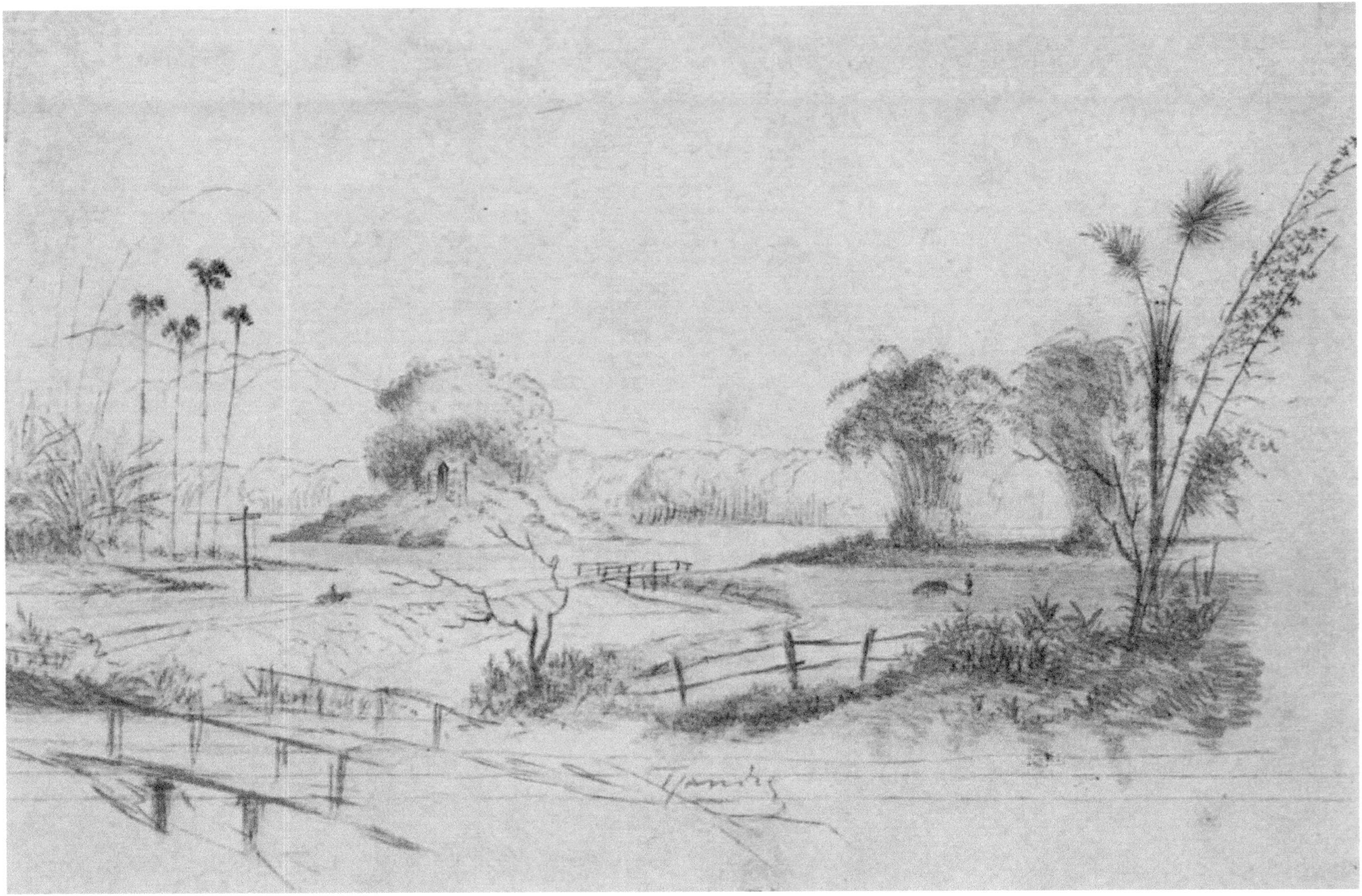

J. D. van Herwerden   Tjandi   K. 10, 21

K. 8, 1—9 Ink drawings. 28,5 × 21,5 cm.

K. 8, 1 Frontispiece to Vol. I: drawing composed of features seen in reliefs of the Candi Jawi: a poet musing; a Javanese gate in the background (p. II, Vol. I).

K. 8, 2 A Javanese book written on palm-leaves, with wooden boards bound with a string (p. XVI, Vol. I).

K. 8, 3 A bird's-eye view of a Royal compound, with various gates and buildings, freely copied from a Trawulan relief (p. 77, Vol. I).

K. 8, 4 A view of the residence of an ecclesiastical gentleman of high rank, drawing composed of features seen in reliefs of the Candi Tigawangi and other monuments of the same period (p. 95, Vol. I).

K. 8, 5 Three villagers carrying fruits and bamboo containers with liquor, suspended from carrying-poles; copied from a relief of the Candi Jawi (p. 125, Vol. I).

K. 8, 6 Frontispiece: A bird's-eye view of a temple compound belonging to a monastery, showing the principal Divine Abode, with a tiered roof and four open pavilions, four-pillared, with a fifth pillar in the centre. The whole is enclosed by a high wall. The small pavilion outside the enclosure is a one-pillared shrine used for religious ceremonies probably connected with the worship of quardian spirits. Copied from a stone relief of the Candi Jago (p. II, Vol. II).

K. 8, 7 A prince seated on a low dais in a park in the company of one attendant, holding in his right hand a token of his master's dignity, probably a cuspidor and a pair of dogs. The prince is receiving gifts of homage, probably food, offered by six men who hold it on their raised right hands. The men are ceremonially introduced by a speaker who is making the añjali salutation in front of the prince. Behind them are seated eight gentlemen of rank in two rows, accompanied by two servants with their hair in pigtails, evidenly prototypes of the *panakawans* (Sĕmar c.s.) of the Javanese wayang theatre. The six men and the speaker are kneeling in a posture of reverence; the others are seated on the ground, talking together quite freely. Copied from a stone relief of the Candi Jawi (p. 108, Vol. II).

K. 8, 8 Procession of a bride, probably a princess, carried by two men in a palanquin, followed by two women and preceded by a young man carrying a state sunshade. The procession is headed by an ordained religious woman of high rank acting as a guardian. She has a sacerdotal headdress and she wears a long robe covering the breasts. She is supervising the two young girls of rank who walk in front of her, acting as bridesmaids (*patah*). Two commoners who are looking on, put their fists into their mouths as an expression of amazement and admiration. The dog walking near the palanquin is an auspicious companion; probably that is the reason why it is given such a prominent place. Copied from a stone relief on the *pĕndapa* terrace of the Candi Panataran (p. 132, Vol. II).

K. 8, 9 Five commoners, under the supervision of their master, a trader, riding on horseback. They are conveying wooden planks to be used as building material which are loaded on a cart and on a bullock's neck. The animal is rebelling against the too heavy load. Freely copied from an old Balinese painting from Klungklung, South Bali, illustrating a tale from the Tantri, the old Javanese version of the Indian Pañcatantra (p. 153, Vol. II).

GEELHOED, J. J.

Approx. 1855.

The manuscripts H 400a and b, library of the Institute, "Het eiland Banka en zijne uitwendige gesteldheid, opgedragen aan Z.E. Generaal Jonkheer F. V. A. Ridder de Stuers Commandant van het O.I. Leger" (The Isle of Banka and its physical appearance, dedicated to His Ex. General Jonkheer F. V. A. Ridder de Stuers, commander in chief of the East Indies army) by L. Ullmann, contains the following water-colours and one pencil drawing, several of which are signed J. J. Geelhoed fecit. Many of the water-colours are touched up with pencil.

H 400a Vol. I, 1855

a. *Orang Darat.* Water-colours. 32 × 20 cm.

b. *Singkee* (Singkeh). Water-colours. 32 × 20 cm.

c. *Kollong mijn* (Kolong mine). Water-colours. 20 × 32 cm.

d. *De Kollong-mijn met de bestanddeelen van het waterschep-rad* (The Kolong mine with component parts of the scoop-wheel). Water-colours. 20 × 32 cm.

e. *Poepoet (Banka)* (Puput (smelting works) Banka). Water-colours. 19,5 × 29,5 cm.

H 400b Vol. II, 1855

a. *Gunung Maras with mouth of Sungei Layang.* Water-colours. 21 × 32 cm.

b. *Boerong Siöol* (Burung Siul). Water-colours. 21 × 32 cm.

c. *Boerong Poeney, Boerong Totes* (Burung Punay and Burung Totes (Authreptes Malacensis)). Water-colours. 21 × 32 cm.

d. *Boerong Beo en Boerong Sernidit* (Burung Beo and Burung Sernidit (Bettet: Conurus alexandri)). Water-colours. 32 × 21 cm.

e. *Boerong Djemboeng, Aijam Hoetan* (Burung Djembung, Ayam Hutan (Caloperdix oculed?)). Water-colours. 21 × 32 cm.

f. *Boewa Djarang* (Buah Djarang (Arrabidaea magnifica?)). Water-colours. 32 × 21 cm.

g. *Potat.* Water-colours. 32 × 21 cm.

h. *Rengas en Amplas* (Rengas (Gluta Renkas) and Amplas). Water-colours. 32 × 21 cm.

i. *Flower* (unfinished). Water-colours. 32 × 21 cm.

j. *Krandang* (Canavalia ensiformis or Puerdria phaseoloicles). Water-colours. 21 × 32 cm.

k. *Timon-dorian*. Water-colours. 32 × 21 cm.

l. *Njereh*. Water-colours. 32 × 21 cm.

m. *Srikaja* (Annòna squamosa L.). Water-colours. 32 × 21 cm.

n. *Bentaro*. Water-colours. 32 × 21 cm.

o. *Retah Maas*. Water-colours. 21 × 32 cm.

p. *Kayoe-ara en Kayoe-akkar (klimoprank)* (Kayu ara and Kayu akkar (creeper)). Water-colours. 21 × 32 cm.

q. *Roemah Darat* (Rumah Darat (pile-dwelling)). Water-colours. 21 × 32 cm.

r. *Batoe Balik (nabij Muntok)* (Batu Baleh (near Muntok)). Water-colours. Signed J. J. Geelhoed. Dated 1854. 21 × 32 cm.

s. *Kg. Boekiet Bt. Poeney Bt. Djeloetoe Gg Maras bezaar en ketjil* (Kampong Bukit, Mount Punai, Mount Djelutu, large and small Mount Maras). Water-colours. 21 × 32 cm.

t. *Tandjg. Raya Banka*. Water-colours. 21 × 32 cm.

u. *Kampg. Broeas aan den voet van de Gg. Koekoes en Langier* (Kampong Bruas at the foot of Mount Kukus and Langir). Pencil. 21 × 32 cm.

v. *Kampong Permis aan den voet van het Permissan gebergte* (Kampong Permis at the foot of the Permissan mountains). Water-colours. Signed J. J. Geelhoed. Dated 1854. 21 × 32 cm.

w. *Tandj. Brikat* (Tanjung Berikat). Water-colours. 21 × 32 cm.

x. *Fort Toboaly*. Water-colours. 17 × 25 cm.

y. *Batoe Praauw* (Batu Proa (seen from two sides)). Water-colours. 32 × 21 cm.

K. 9, 1 *Het Fort Toboalij gezien van de landzijde* (View of Fort Toboaly from the landward side).
Water-colours. Signed J. J. Geelhoed fec. Dated 1859 (the 9 has been superimposed on another figure). 18 × 31 cm.
Note: The drawing is somewhat different from H 400b No. x.

Provenance: The manuscripts H 400a and 400b were presented by A. L. de Sturler, Leiden, Febr. 15, 1896. No. K. 9, 1 was bought from the antiquary Loose, The Hague.

GRIJZEN, H. J.

Approx. 1892. Referred to as 'Aspirant-Controleur' (lowest rank in civil service) in the districts of Mana and Pasemah (*Regeerings Almanak*, 1893).
The manuscript H 799 "De Bĕsĕmahers, Volkenkundige Bijdragen" (The people of Bĕsĕmah, Ethnographical Contributions) by O. L. Helfrich contains 9 foolscap sheets on which 23 drawings of betel leaves plaited in various patterns. Blue pencil. 30 × 21 cm. These drawings belong to the chapters on religion and sexual life.

HAMILTON HAMMON, G.

Sydney, Australia.
F. 14, 13 *Djandi Semarang Java* (Tjandi Semarang, Java).
Water-colours on Whatman paper. Signed G. Hamilton Hammon. 36 × 52 cm.
Note: Touched up with pencil, folded in the middle. Mark on upper edge II F. On reverse side: A view of Semarang (taken from D. M. Camparell's house at Tjandi).

HARDOUIN, Ernest

French artist approx. 1850.
Many water-colours and pencil drawings were made by this artist during his stay in Batavia, Indonesia. These drawings were reproduced in the book *Java, tooneelen uit het leven, karakterschetsen en kleederdrachten van Java's bewoners; in afbeeldingen van de natuur geteekend* door E. Hardouin; met tekst van W. L. Ritter en een voorwoord van H. M. Lange, gepensioneerd luitenant-kolonel van het O.I. leger; uitgegeven te 's-Gravenhage bij K. Fuhri. 1855 (Java, scenes taken from life, character sketches and costumes of the inhabitants of Java; pictures drawn from nature by E. Hardouin; text by W. L. Ritter and a preface by H. M. Lange, pensioned Lieut.-Col. of the East Indies Army; published in The Hague by K. Fuhri. 1855). One single water-colour was included in another book by W. L. Ritter, *De Europeaan in N. Indië* (The European in the Netherlands Indies) and also in the picture book *De Indische Archipel* (The Indian Archipelago). The drawings were lithographed in the workshop of Lemercier, Paris. The original water-colours and drawings have been mounted on 33 pieces of pasteboard and filed in a red cloth portfolio with leather back and corners. 42 × 27 cm.

F. 6,1 *Design for the title-page "Java karakterschetsen, klederdrachten, houding en voorkomen; door Ernest Hardouin. Batavia, 1851* (Java, character sketches, costumes, posture and appearance; by Ernest Hardouin. Batavia, 1851).
Sepia, highlighted with white on brown paper. 35,5 × 29 cm.
Note: Title-page of an earlier book which was never published.

F. 6, 2 *Javanese servant.*
Water-colours. Signed Ernest Hardouin. 20 × 15 cm.

F. 6, 3 *Njai or Native housekeeper.*
Water-colours. Signed Ernest Hardouin. 20 × 15 cm.

F. 6, 4 *An opium den.*
Water-colours. Signed Ernest Hardouin. 15 × 19,5 cm.

F. 6, 5 *African soldier.*
Water-colours. Signed Ernest Hardouin. 20 × 14,5 cm.

F. 6, 6 *Batavia Bride.*
Water-colours. Signed Ernest Hardouin. 19,5 × 14,5 cm.

F. 6, 7 *Arab in Java.*
Water-colours. Signed Ernest Hardouin. 20 × 14,5 cm.

F. 6, 8 *Bathing slave.*
Water-colours. Signed Ernest Hardouin. 19,5 × 15 cm.

F. 6, 9 *Javanese man.*
Water-colours. 20 × 15 cm.

F. 6, 10 *Javanese woman.*
Water-colours. Signed Ernest Hardouin. 19,5 × 15 cm.

F. 6, 11 *Javanese priest.*
Water-colours. Signed Ernest Hardouin. 20 × 15 cm.

F. 6, 12 *Kong-a-lian player.*
Water-colours. Signed Ernest Hardouin. 19,5 × 15 cm.

F. 6, 13 *Regent in Java.*
Water-colours. Signed Ernest Hardouin. 19 × 15 cm.

F. 6, 14 *Topeng Babakan.*
Water-colours. 19,5 × 14,5 cm.
Note: The drawing shows mask-dancers.

F. 6, 15 *Chinese priest of Batavia.*
Water-colours. Signed Ernest Hardouin. 19,5 × 14,5 cm.

F. 6, 16 *Coolies.*
Water-colours. Signed E. Hardouin. 19,5 × 15 cm.

F. 6, 17 *A Javanese nobleman in hunting attire.*
Water-colours. Signed E. Hardouin. 19,5 × 14,5 cm.

F. 6, 18 *Bridegroom from the Sunda countries (Java).*
Water-colours. Signed E. Hardouin. 20 × 15 cm.

F. 6, 19 *Native woman from the interior of Palembang.*
Water-colours. Signed Ernest Hardouin. 19,5 × 14,5 cm.

F. 6, 20 *Dancing-girl at a Javanese court.*
Water-colours. Signed E. Hardouin. 20 × 15 cm.

F. 6, 21 *Chicken merchant.*
Water-colours. 20 × 15 cm.

F. 6, 22 *Javanese sailor.*
Water-colours. 24 × 15,5 cm.
Note: Mirror image of plate in the book.

F. 6, 23 *Chained convict.*
Water-colours. 23 × 15 cm.
Note: Mirror image of plate in the book.

F. 6, 24 *Javanese street-dancer.*
Water-colours. Signed Ernest Hardouin. 22 × 14 cm.

F. 6, 25 *Javanese Ronggeng.*
Water-colours. Signed Ernest Hardouin. 22 × 14 cm.
Note: The drawing shows a dancing girl.

F. 6, 26 *Chinese smith, Java.*
Water-colours. 23,5 × 15,5 cm.

F. 6, 27 *Djayang Sekar.*

Water-colours. 23,5 × 15,5 cm.
Note: The drawing shows a Javanese cavalryman.

F. 6, 28 *European gentleman in Java.*
Water-colours. 23,5 × 15,5 cm.
Note: European gentleman (white suite and black top hat). Reproduced in *De Europeaan in Ned. Indië* by W. L. Ritter.

F. 6, 29 *Sailor (pirate?).*
Ink and water-colours. Signed Ernest Hardouin. 21,5 × 17,5 cm.
Note: Not reproduced in *Java, tooneelen uit het Leven,* etc., by W. L. Ritter. Perhaps a preliminary study for F. 6, 22. On the reverse side a small pencil sketch of Muslim grave.

Mounted on 4 pieces of pasteboard are 14 small letter-illuminations, three of which have not been reproduced in *Java, tooneelen uit het leven,* etc. All vignettes are pencil drawings.

F. 6, 30 *K. Landing native sloop.*
9 × 13 cm. Square.

F. 6, 31 *N. Stamping rice.*
Signed E. H. 9 × 9 cm. Edges cut off.

F. 6, 32 *P. Native shop.*
Signed E. H. 8,5 × 8,5 cm. Edges cut off.

F. 6, 33 *Y. Waterfall.*
Signed E. H. 8,5 × 7 cm. Edges cut off.

F. 6, 34 *X. Woodland scene.*
Signed E. H. 7 × 8,5 cm. Edges cut off.

F. 6, 35 *B. Opium-smoker.*
Signed E. Hardouin. 9,5 × 8,5 cm. Square.

F. 6, 36 *D. Groom.*
9,5 × 9 cm. Square.

F. 6, 37 *E. Native woman with child.*
12 × 8 cm. Square.

F. 6, 38 *W.*
9,5 × 9,5 cm. Square.

F. 6, 39 *O. Native angler.*
Signed Ernest Hardouin. 9,5 × 9,5 cm. Square.

F. 6, 40 *J. Native cavalryman.*
Signed E. H. 8 × 7,5 cm. Edges cut off.

F. 6, 41 *Chinese family on a horse-drawn cart.*
Signed Ernest Hardouin. 10,5 × 19 cm. Edges cut off.

F. 6, 42 *H. Mosque, Batavia.*
Signed E. H. 8 × 11,5 cm. Edges cut off.

F. 6, 43 *G. European hunter.*
9 × 9 cm. Edges cut off.

Provenance: Bought at Van Pelt's antiquary, The Hague. 1962.

HARMSEN VAN DER BEEK, Eelco M. ten

Approx. 1913.

F. 7, 1—65 *Javanese krisses, klewangs, daggers.*
Water-colours, the damascene (*pamor*) of the krisses has been rendered with silver. Some of these paintings are signed Eelco M. ten Harmsen van der Beek A'dam '13. All of the 65 drawings 50 × 32,5 cm., in portefolio 54 × 37 cm.
Note: *Pamor* is the result of mixing meteoric iron with ordinary iron. The weapons depicted are not only from Java but also from Bali and even from the Dayak countries.

Provenance: Purchased in 1958 from Van der Peet's antiquary, Amsterdam.

HASSELT, Arend Ludolf van

Born Jan. 6, 1848, Groningen. Died Sept. 29, 1909, Oosterbeek.
He began his career as a sergeant in the infantry and ended it as a professor of Ethnology at the Indische Instelling (Indies Institute) in Delft. In 1873 'controleur' 2nd class in the West Coast of Sumatra Residency; obtained permission to join an expedition to Central Sumatra organised by the Netherl. Geographical Society. He was in charge of the non-nautical section. From 1879—1882 he was in Holland to work out the data collected during the expedition. He made several drawings for the book *Midden-Sumatra* (Leiden 1881—92) which he wrote in collaboration with some other members of the expedition. The 17 small water-colours following have probably also been painted by him. They have been drawn on the reverse side of visiting cards, the majority of them cards of officials in office in the West Coast of Sumatra Residency roundabout 1874. Four of these are A. L. van Hasselt's own (Nos. 7, 9, 11, and 16). All of these are mounted, approx. 16 × 19,5 cm.

O. 3, 1 *Zeegezicht bij Padang* (Sea-view near Padang).
5,4 × 9,5 cm.

O. 3, 2 *Danau di Atas* (Upper lake).
5 × 8,5 cm.

O. 3, 3 *Rijstschuur te Soepajang* (Rice-granary at Supajang).
5 × 8,6 cm.

O. 3, 4 *Missigit te Soepajang* (Mosque at Supajang).
5,2 × 9,4 cm.

O. 3, 5 *Woning te Soepajang* (House at Supajang).
5,4 × 9,5 cm.

O. 3, 6 *Meer van Singkarah* (Lake Singkarah).
5 × 8,8 cm.

O. 3, 7 *Regenbui op het meer van Singkarah* (Shower on Lake Singkarah).
5,6 × 9,5 cm.

O. 3, 8 *Meer van Singkarah* (Lake Singkarah).
5 × 8,8 cm.

O. 3, 9 *Meer van Singkarah* (Lake Singkarah).
5,5 × 9,5 cm.

O. 3, 10 *Danau di Atas* (Upper lake).
5 × 8,7 cm.

O. 3, 11 *Danau di Atas* (Upper lake).
5,5 × 9,5 cm.

O. 3, 12 *Waterval in de kloof van de Batang Anei* (Waterfall in the gorge of the Batang Anei).
8,8 × 5,2 cm.

O. 3, 13 *Waterval bij Boengoes* (Waterfall near Bungus).
8,8 × 5,3 cm.

O. 3, 14 *Zeegezicht bij Padang* (Sea-view near Padang).
5 × 8,3 cm.

O. 3, 15 *Batang lawé bij Soepajang* (Batang Lawé near Supajang).
5 × 9,5 cm.

O. 3, 16 *Meer van Singkarah* (Lake Singkarah).
5,5 × 9,5 cm.

O. 3, 17 *Vlakte van Solok* (Plain of Solok).
4,9 × 8,3 cm.

Provenance: Purchased at Loose's antiquary, The Hague, 1964.

HERWERDEN, Jacob Dirk van

Born 1806. Died 1878.

Went to the Neth. Indies as an ordinary seaman but switched over to the civil service. In 1834 he was 'controleur' of Ngrawa. From 1835—1840 assistant resident in the same district. From 1841—1844 held same function at Prabalingga. From 1844—1848 resident at Madiun. He made many sketches in water-colours and pencil. After repatriation he made oil-paintings based on these sketches. Sensitive work, highly esteemed by dr. Corn. Hofstede de Groot. Herwerden was a pupil of the Belgian landscape painter Roffiaen. He has exhibited his work several times.

S. 8 *Kampong Magettan.*
Oil. Signed J. D. v. Herwerden. 55 × 70 cm. In gilt frame.
Note: Magetan near Madiun.

S. 9 *Tenggereesch dorp Masangan bij Wonokitri* (Masangan, village in the Tengger near Wonokitri).
Oil. Signed J. D. v. Herwerden. 53 × 82 cm. In gilt frame.

S. 10 *Soekapoera* (in the Tengger).
Oil. Signed J. D. v. Herwerden. 55 × 72 cm. In gilt frame.

K. 10 *Sketch-book.*
86 pages, Old Dutch paper. 31,5 × 20 cm. Contents: many sketches, mostly in pencil, of landscapes, trees, plants, animals, houses, etc., mainly in East Java, with notes on the colours; some of these are dated (from Dec. 6, 1847—March 21, 1848).

K. 10, 1 *Wielis gebergte van Ponorogo te zien* (View of Wilis Mountains from Ponorogo).

K. 10, 2 *Javanese desa with bamboo and palm-trees.*

K. 10, 3 *Various subjects: Squatting Javanese men, bamboo plait-work, young coconut tree, etc.*
Dated '16 December 47' and '3 Jan. 48'.

K. 10, 4 *Empty village market with only one stall carrying on trade.*

K. 10, 5 *Various subjects: Woman carrying pots on her back, a coachman, a Javanese man ploughing, fighting-cock, washing on a line, etc.*
Note: Large number of captions.

K. 10, 6 *Buffalo-carts.*
Dated '18 December [1848]'.

K. 10, 7 *Pinang palm.*
Dated 'van Sarangan 26 januarij 48'.
Note: With detailed description.

K. 10, 8 *Trees.*
Dated '26 januarij 48' and '27 Jan. [1848]'.
Note: With detailed description.

K. 10, 9 *Palm-trees (ordinary pinang palm, pakis hadji tareca).*
Dated '26 Febr. 48'.

K. 10, 10 *Kapok-tree and sketch of a government building (Bogor?).*

K. 10, 11 *Clumps of banana-trees (Musa).*

K. 10, 12 *Clump of bana-trees and Javanese man with hoe.*

K. 10, 13 *Palm-trees, among others a salak-palm.*
Dated 4 febr. 48.

K. 10, 14 *Pandanus-tree mounted on high surface-roots.*
Dated 7 December 1847.

K. 10, 15 *Sketch of mountain landscape with desa (Tengger?).*

K. 10, 16 *Sketch of mountain landscape.*

K. 10, 17 *Village houses.*

K. 10, 18 *Tree outside village, smoking volcano in background.*
Dated '17 Dec. 47'.

K. 10, 19 *Tree at the side of a road running along a lake or the sea.*
Sepia.

K. 10, 20 *River between overgrown rocks.*
Sepia.

K. 10, 21 *Tjandi.*
Note: Landscape in background.

K. 10, 22 *The main road with halting-place.*

K. 10, 23 *Sketch of a bay.*

K. 10, 24 *Sketch of section of a ship's mast.*

K. 10, 25 *Masangan near Wonokitri, Tengger.*
Note: With pencilled grid. Preliminary study for S. 9.

K. 10, 26 *Palms and other trees.*

K. 10, 27 *Road in mountain scenery.*

K. 10, 28 *Mountain landscape with sawahs.*

K. 10, 29 *Sketch of mountain landscape.*
Note: Marked 'no. 5'.

K. 10, 30 *Solo rivier* (Solo River).
Dated '21 Maart 48'.
Note: With notes and drawings of proas etc.

K. 10, 31 *Sketch of mountain village.*

K. 10, 32 *Sketch of European buildings.*
Note: Also in K. 10, 33.

K. 10, 33 *Village in the Tengger with government buildings (Tosari?).*

K. 10, 34 *Tjemoro Sewoe.*

K. 10, 35 *Mountain-road.*
Ink and pencil.

K. 10, 36 *Road between a village and a river.*

K. 10, 37 *Mountain landscape with volcano.*
Pen and pencil.

K. 10, 38 *St. Helena.*
Water-colours. To be dated 1848.

K. 10, 39, 40 Some detached sheets.

K. 10, 39 *Loemadjang.*
Pencil. Dated 1829. 17,5 × 27 cm.
Note: The drawing shows a civil servant's house with flag.

K. 10, 40 *Ngadie Poerie.*
Pencil. Dated 11 Sept. 1841. 13 × 7 cm.
Note: The drawing shows a Hindu-Javanese tempel.

Provenance: Legacy 1878.

HEYDEN, J. van der

He probably arrived in the Indies about 1908 and died at Meester Cornelis in 1928. In 1920 he is mentioned as a decorator and in 1927 as a painter. He probably was the author of a booklet *Batavia bij nacht* (Batavia at night), containing a somewhat fantastic biography by M. B. Coelho, dated '25 April 1928', 2nd edition '30 Maart 1940'. He is chiefly known from his rather stiff book-illustrations and pictures for class-rooms. His earliest known work: J. van der Heyden, *Indisch prentenboek. I. Bedienden en beroepen* (Indies picture-book. I. Servants and trades), Amsterdam, J. H. de Bussy, [1909]. Later appeared *Nieuw Indisch A.B.C.* (New Indies A.B.C.) with Dutch and Malay text, published by Visser and Co. in Batavia. Since 1918 he worked for Volkslectuur (Balai Pustaka), illustrated a.o. the Hikayat Pandji Semirang, Mh. Roesli's Siti Noerbaja. The firm of Kolff commissioned him to illustrate *Loetoeng Kasaroeng,* a Sundanese legend freely re-written by F. A. Schöppel, Batavia 1925. On page 3 of the book *Oud Soerabaia* (Old Surabaya) by G. H. von Faber (Soerabaja 1931) is a reproduction of a picture, the product of a very fanciful imagination, of Hayam Wuruk's pageant. In this book is also mentioned the year of Van der Heyden's death.

O. 4 *Illustrations for a school-reader.*
In 3 vols., each of 20 lessons. The first volume deals with the school-days of a Javanese boy. The second volume is about life at a sugar refinery, where the boy is a servant of a European family. The third volume is about the life of the same family after moving to Surabaya. The numbers of the lessons to which the illustrations belong are given on the reverse side. Some of the drawings have been mounted, so that the numbers are not visible, but these can be placed approximately. All the drawings are 14,5 × 19 cm.

O. 4, 1 *A Javanese boy with slate and book on the way to school, in front of a batik-shop.*
Water-colours. Signed J. van der Heyden.
Note: On reverse side in pencil: 1e stukje 1e les (vol. 1, 1st lesson). Numbered 1/1.

O. 4, 2 *Five Javanese boys at a school-desk.*
Sepia. Signed J. van der Heyden.
Note: The first boy is the same as in O. 4, 1. Numbered 1/2.

O. 4, 3 *Batik-shop with warong next to it.*
Sepia. Signed J. van der Heyden.
Note: The same shop as in O. 4, 1.

O. 4, 4 *Woman and girl in a batik-shop.*
Sepia. Signed J. van der Heyden.
Note: The same shop as O. 4, 1. Numbered 1/5.

O. 4, 5 *Girl with a small child in front of a batik-shop.*
Sepia. Signed J. van der Heyden.
Note: The same shop as in O. 4, 1, the same girl as in O. 4, 4. Numbered 1/6.

O. 4, 6 *Boy in front of a closed batik-shop looking at a warong.*
Water-colours. Signed J. van der Heyden.
Note: The same shop as in O. 4, 3, the same boy as in O. 4, 1, 2. Numbered 1/7.

O. 4, 7 *Two boys near their school.*
Sepia. Signed J. van der Heyden.
Note: Numbered 1/8.

O. 4, 8 *Five boys on the school verandah, a European gentleman (the teacher?) is approaching.*
Sepia.
Note: Numbered 1/9.

O. 4, 9 *Two boys, one of whom is flying a kite.*
Sepia. Signed J. v. d. H.
Note: Numbered 1/14.

O. 4, 10 *The boy (ill?) in bed, his kite is hanging on the wall, a woman (his mother?) is giving him something to drink.*
Water-colours. Signed J. v. d. H.
Note: Numbered 1/16.

O. 4, 11 *Inside batik-shop. One woman is showing a kain to two other women.*
Sepia. Signed J. v. d. H.
Note: Numbered 1/17.

O. 4, 12 *The boy, with a payong over his head, arriving at school in the rain.*
Sepia. Signed J. v. d. H.
Note: Numbered 1/18.

O. 4, 13 *A man looking at a kain in the batik-shop.*
Water-colours. Signed J. van der Heyden.

O. 4, 14 *The boy, a woman with a child and the coachman in a sado (horse-drawn carriage), a girl is standing behind the sado, in front of the batik-shop.*
Sepia. Signed J. v. d. H.
Note: Numbered 1/20.

O. 4, 15 *Sugar-factory, grobak (cart) with sugar-cane in foreground.*
Sepia. Signed J. v. d. H.
Note: Numbered 2/2.

O. 4, 16—17 *Grobaks (carts) with sugar-cane in the fields.*
Sepia.
Note: The two drawings are the same. One signed J. v. d. H. Numbered 2/3.

O. 4, 18 *Unloading sugar-cane at the factory.*
Water-colours. Signed J. v. d. H.
Note: Numbered 2/4a.

O. 4, 19—20 *Crushing of the cane.*
Sepia.
Note: The two drawings are the same. Numbered 2/4b.

O. 4, 21 *Overseer and coolie (the boy?) near a native house.*
Water-colours. Signed J. v. d. H.
Note: Numbered 2/5.

O. 4, 22—23 *The boy (?) strolling along a path, a jungle-fowl in the foreground.*
Sepia.
Note: The two drawings are the same. One signed J. v. d. H. Numbered 2/6.

O. 4, 24 *Man with hoe in his hand looking at a woman gathering corn.*
Sepia. Signed J. v. d. H.
Note: Numbered 2/7.

O. 4, 25 *The man with the hoe walking from the cornfield to a bamboo house, followed by a man with a walking-stick.*
Water-colours. Signed J. v. d. H.
Note: Numbered 2/8?

O. 4, 26 *The sugar-refinery at night.*
Sepia. Signed P. v. T.
Note: Numbered 2/9.

O. 4, 27 *The European administrator (here with bald head and a large dark moustache) in his office. An employer (Indo-European), coming from the fields, is joining him.*
Sepia. Signed J. v. d. H.
Note: Numbered 2/10.

O. 4, 28 *An overseer sitting on a bunk, writing; in front of him are 4 coolies squatting on the ground.*
Sepia. Signed J. v. d. H.
Note: Numbered 2/11.

O. 4, 29 *Shop, lighted by an oil-lamp.*
Sepia. Signed J. v. d. Heyden.
Note: Numbered 2/12.

O. 4, 30 *Javanese wedding.*
Water-colours. Signed J. v. d. H.
Note: Wedding of the man and the woman of O. 4, 24?

O. 4, 31 *Two men, with baskets hanging from yokes, arriving at a bamboo house, in front of which are a woman, a boy, and a girl.*
Sepia. Signed J. v. d. H.
Note: Numbered 2/14.

O. 4, 32 *Two men refreshing themselves at a pool or well.*
Water-colours. Signed J. v. d. H.
Note: Numbered 2/15.

O. 4, 33 *An ayah with a European child in her slendang and a European boy on the steps of the front verandah. The father is arriving in a bandy.*
Sepia. Signed J. v. d. H.
Note: Numbered 2/16.

O. 4, 34 *The wife of the administrator giving something to the house-boy while her small son is feeding the ducks.*
Sepia. Signed J. v. d. H.
Note: Numbered 2/17.

O. 4, 35 *The administrator with a Javanese civil servant, a native policeman, and a coolie near a broken-down fence along the river.*
Water-colours. Signed J. v. d. H.

O. 4, 36 *The administrator arriving home, where breakfast is prepared; a lady is cutting bread, another lady and a small son are playing with building-blocks.*
Sepia. Signed J. v. d. H.
Note: Numbered 2/19.

O. 4, 37 *The family leaving in a carriage (for Surabaya?).*
Sepia. Signed J. v. d. H.
Note: Numbered 2/20.

O. 4, 38 *European house (in Surabaya?).*
Chalk. Signed J. v. d. H.
Note: Numbered 3/2.

O. 4, 39 *Lady in front verandah. A boy in night-gown is mounting the steps.*
Chalk. Signed J. v. d. H.
Note: The same house as O. 4, 39. Numbered 3/3.

O. 4, 40—41 *The family by the oil-lamp at night.*
Sepia.
Note: The two drawings are the same. One signed J. v. d. H. Numbered 3/5.

O. 4, 42 *Lady of the house, gardener, house-boy, and ayah.*
Sepia. Signed J. v. d. H.
Note: Numbered 3/6.

O. 4, 43 *Merry-go-round with children.*
Sepia. Signed J. v. d. H.
Note: Numbered 3/8.

O. 4, 44 *Perak (port of Surabaya).*
Water-colours.

O. 4, 45 *Ships in the roads (of Surabaya).*
Sepia.
Note: Numbered 3/11.

O. 4, 46 *Mother at sewing-machine reading a letter to two boys and a girl.*
Sepia. Signed J. v. d. H.
Note: Numbered 3/12.

O. 4, 47—48 *Girl looking after the plants, helped by the gardener.*
Water-colours.
Note: The two drawings are the same. One signed J. v. d. H. Numbered 3/13.

O. 4, 49 *Mother sewing while two small children are playing on the ground. Ayah in the background.*
Sepia.
Note: Numbered 3/14.

O. 4, 50 *The children's father has come for a visit and is greeted by the three elder ones coming home from school.*
Sepia. Signed J. v. d. H.
Note: Numbered 3/15.

O. 4, 51 *Three European boys and a Javanese boy looking down from a bridge at a bandjir (river-flood).*
Sepia.
Note: Numbered 3/17.

O. 4, 52 *Boats in flooded river.*
Sepia. Signed J. v. d. H.
Note: Numbered 3/18.

O. 4, 53—54 *Sick girl in a reclining chair. The Javanese boy is bringing flowers, while her mother is pouring out some medicine.*
Water-colours.
Note: The two drawings are the same. One signed J. v. d. H. Numbered 3/19.

O. 4, 55 *The girl convalescing in reclining chair. Her father (administrator) and mother, the two European boys and the Javanese boy (modestly keeping himself in the background) are sitting with her.*
Sepia. Signed J. v. d. H.
Note: Numbered 3/20.

O. 4, 56 *One of the two European boys on a bicycle.*
Water-colours. Signed J. v. d. H.

O. 4, 57 *Concordia, Sociëteitsstraat* (a street in Surabaya).
Water-colours. Signed J. v. d. H.

O. 4, 58 *Kajoen* (street in Surabaya).
Water-colours. Signed J. v. d. H.

Provenance: Bought at Loose's antiquary, The Hague, 1963.

HILHORST, G. P. L.

Born March 19, 1886, Hillegersberg (near Rotterdam).
He trained at the State School for Industrial Arts in Amsterdam (1902—1905), pupil of Lion Cachet. Went to the Neth. Indies in 1920 as an art teacher. Lived there for 19 years. Among his works are many ex-libris.

GF. 1, 9 *Flight of stairs at Borobudur.*
Wood-cut. Signed G. P. L. Hilhorst 16/25 Borobudur. 41,5 × 28,5 cm.
Note: In upper right corner: MCMXXI GPLH.

Provenance: Bought at Loose's antiquary, The Hague, 1963.
References: J. Schwenke, Exlibriskunst, *Cultureel Indië,* Vol. 5, pp. 165—169.

HOPPENSTOK, H. B.

Approx. 1816.

O. 5 Album (32 pp.; 20 × 16,5 cm) containing 22 ink drawings; modern binding. According to a note (by B. van Suchtelen, 11th December, 1940) attached to the fly-leaf, one leaf with a note giving the name of the artist and the date 1816, and one with a note according to which the collection was bought at an auction at Van Stockum's in Nov. 1903, was lost by the binder. Presumably, the drawings may be traced to earlier copies. Most of them contain captions in French, German, Dutch, English, and Spanish. A corrected version of the English captions is given in the description below. Where no English caption has been given, the original French or Dutch captions have been retained.

O. 5, 1 *Capture of the town of Den Briel by the Count Van der Mark on behalf of the Prince of Orange on the first of April, 1572.*
Ink. 8,5 × 14,5 cm.

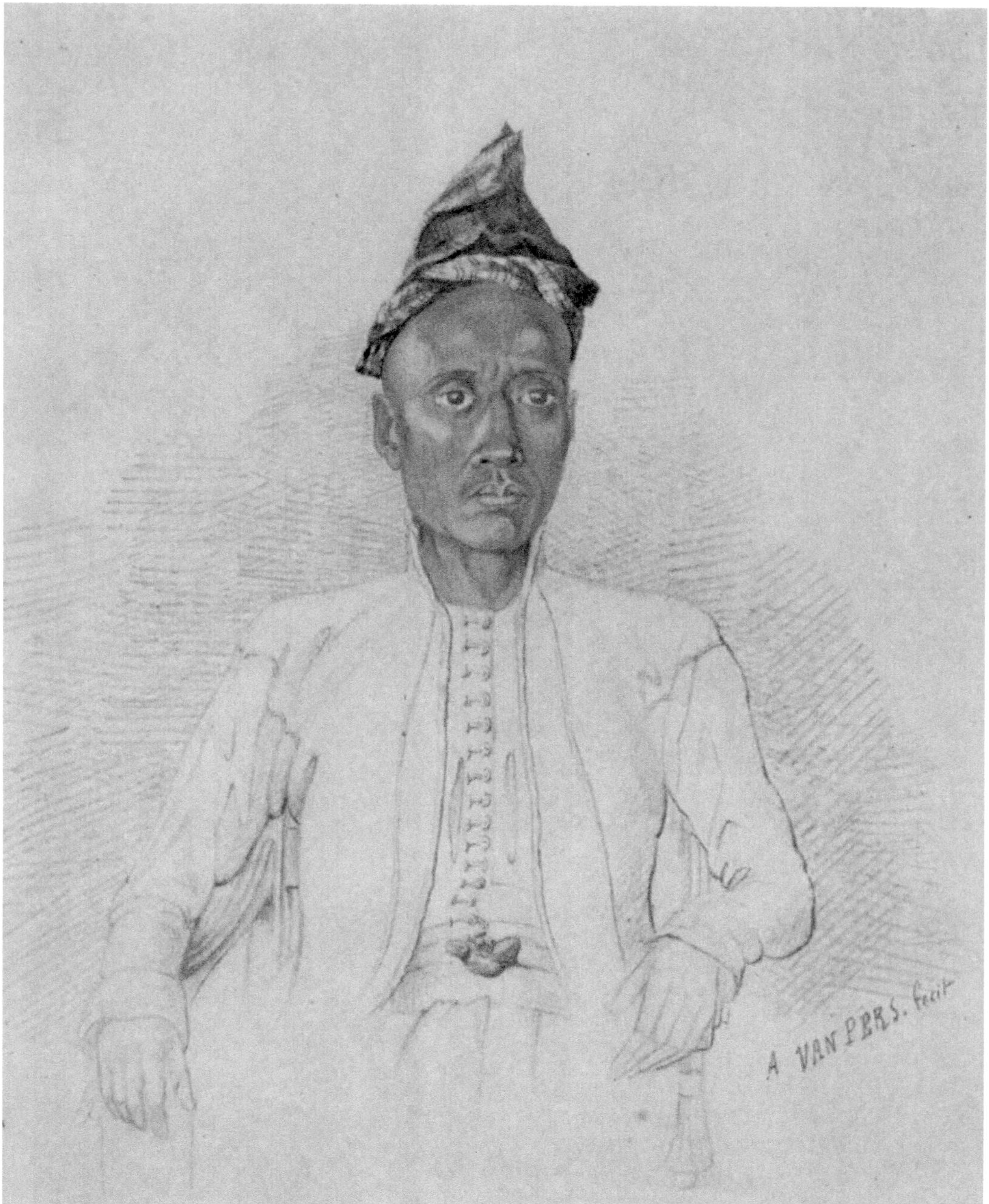

A. van Pers Wealthy Buginese K. 15, 4

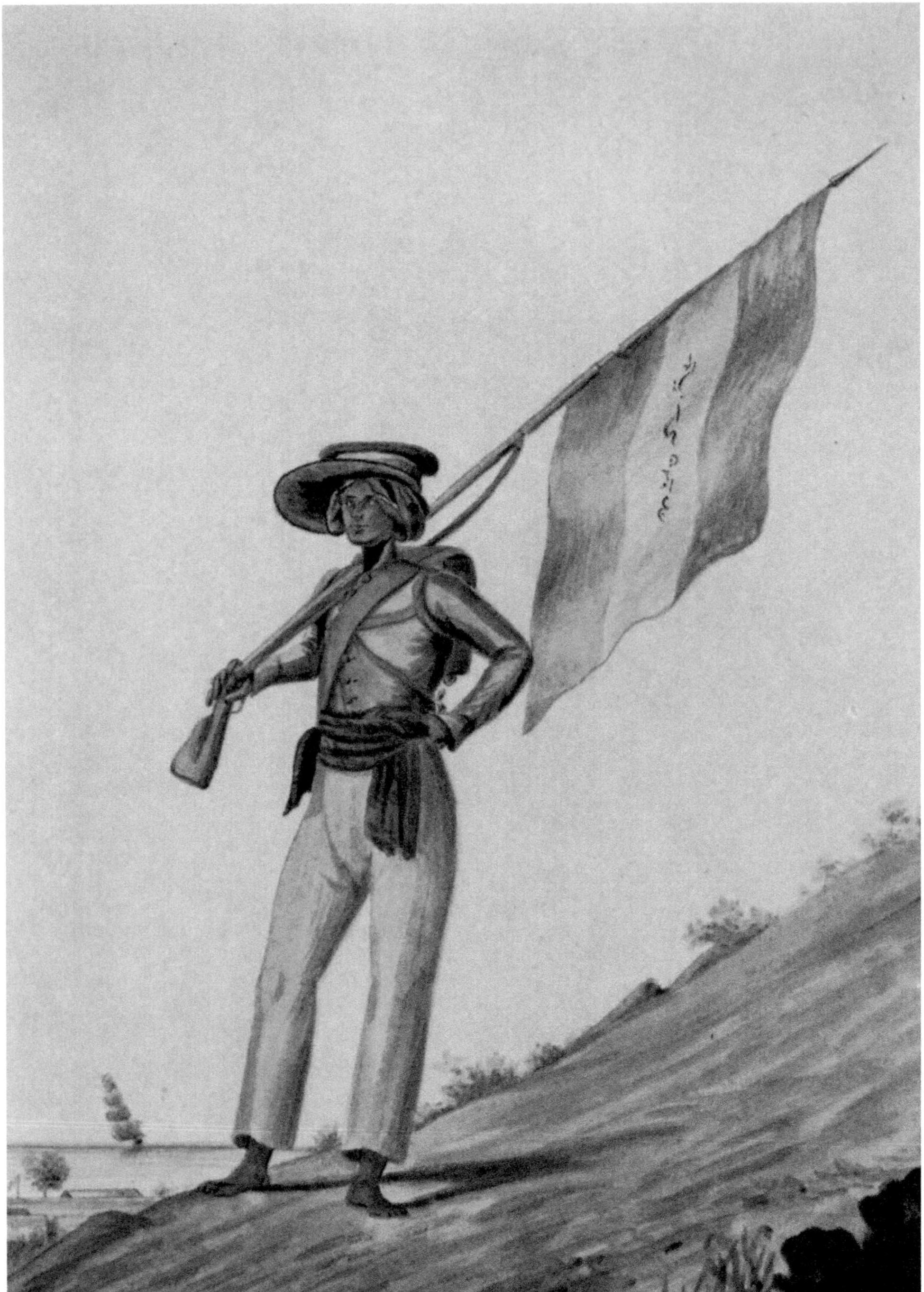

Javanese corporal K. 26, 1; *see* p. 85

O. 5, 2 *The victory of the North-Holland fleet over the Spanish fleet commanded by the Count De Bossu, on the 11th. and 12 th. October, 1573, near the town of Hoorn.*
Ink. 8 × 15 cm.

O. 5, 3 *Herring-fishing.*
Ink. 8 × 15 cm.

O. 5, 4 *The Sunda Straits between the islands of Sumatra and Java.*
Ink. 8,5 × 14,5 cm.

O. 5, 5 *The city of Cochin on the coast of Malabar.*
Ink. 8,5 × 14,5 cm.

O. 5, 6 *L'Isle d'Ormus, Le fort, L'Isle de Larcke, Une partie de l'Isle de Kismus. Les Isles sont situées a l'entrée du Golfe Persique* (The isle of Ormus. The fort. Isle of Larcke. Part of the isle of Kismus. The islands are situated in the entrance to the Persian Gulf).
Ink. 8 × 15 cm.

O. 5, 7 *Vue de Pon. Pandjang, ou het Lang Eiland, den Blaauwen Berg, La baije de Bantam, le Coin de Bantam, au Detroit de la Sonde* (View of Pulu Pandjang, or Long Island, the Blue Mountain, the Bay of Bantam, Bantam Corner, in Sunda Straits).
Ink. 9 × 14,5 cm.

O. 5, 8 *L'Isle Dwars in den Weg, L'Isle Selebese, le continent, au detroit de La Sonde entre les Isles de Sumatra et de Java* (The island of Dwars in den Weg, the island of Sebesi, the mainland, in the Sunda Straits between the islands of Sumatra and Java).
Ink. 8,5 × 14,5 cm.

O. 5, 9 *The island of Kuiper near the island of Onrust opposite Batavia.*
Ink. 8 × 14 cm.

O. 5, 10 *The city of Bantam on the island of Java.*
Ink. 8,5 × 14,5 cm.

O. 5, 11 *La ville d'Archangel sur la Rive droite de la Rivière Dwina* (The town of Archangel on the right bank of the river Dwina).
Ink. 8,5 × 15,5 cm.

O. 5, 12 *Samojeedse Slede met hare Rendieren. Samojeedse Tent. Samojeedse Tent van binnen. Beer. Wolf. Vos. Bever. Veelvraat. Sabelmarter. Hermelijn. Muskus Dier* (Samoyedic sledge with reindeers. Samoyedic tent. Interior of Samoyedic tent. Bear. Wolf. Fox. Beaver. Carcajou. Sable. Ermine. Musk-deer).
Ink. 15 × 16 cm.

O. 5, 13 *The very high hill called Adams Peak upon the island of Ceylon at the entrance of the Bay of Punto Gale.*
Ink. 9 × 14 cm.

O. 5, 14 *Olyfant. Rhinoceros (eenhoornige). Dromedaris. Kameel. Civet kat* (Elephant. Rhinoceros (one-horned). Dromedary. Camel. Civet-cat).
Ink. 13 × 15 cm.

O. 5, 15 *The Greenland fishing-fleet.*
Ink. 8,5 × 14,5 cm.

O. 5, 16 *Walrus. Robbe. Walvis. Ysbeer. Robbe met de witte Buik* (Walrus. Seal. Whale. Polar bear. White-bellied seal).
Ink. 7 × 16 cm.

O. 5, 17 *The island of Onrust opposite Batavia.*
Ink. 8 × 14 cm.

O. 5, 18 *Orang Outang. Baviaan. Wanderoe. Krokodil. Jakhals. Buffel. Betelblad. Arek Vrugt. Arek pit in Betelblad gerold* (Orang-outang. Baboon. Wanderoo. Crocodile. Jackal. Buffalo. Betel leaf. Areca fruit. Stone of the Areca fruit rolled in Betel leaf).
Ink. 12,5 × 15 cm.

O. 5, 19 *The Cape of Good Hope, provisioning port for the East Indies-bound ships.*
Ink. 8,5 × 14,5 cm.

O. 5, 20 *Leeuw. Tyger. Panter. Luipaard. Patas. Gevlekte Hijena. Zebra. Hippopotamus. Giraffe. Koemelken der Hottentotten. Twee-hoornige Rhinoceros* (Lion. Tiger. Panther. Leopard. Patas (monkey). Spotted Hyena. Zebra. Hippopotamos. Giraffe. Hottentots milking cows. Two-horned Rhinoceros).
Ink. 19 × 16 cm.

O. 5, 21 *The City of Batavia on the island of Java.*
Ink. 8,5 × 14,5 cm.

O. 5, 22 *Branche du Café. Branche du Poivre. Clous de Girofle. Le cacao. L'indigo* (Branch of the coffee shrub. Branch of the pepper tree. Cloves. Cocoa. Indigo).
Ink. 18 × 15,5 cm.
Note: With detailed French texts.

ISINGS, J. H.

Born July 31, 1884, Amsterdam.

Pupil of the Art School for Artistic Trades under the direction of J. Visser Jr. and Georg Rueter and later N. v. d. Waay and P. Josselin de Jong. Illustrated i.a.: historical books by Joh. H. Been, two children's bibles (1917—1922), several anthologies for teaching purposes, books on history, *Uit 't sagenland* (From the land of saga's) by N. van Hichtum and Cor Bruyn, *Vaderlandsche geschiedenis* (Dutch National History) by P. Louwerse, 60 biblical pictures (published by Wolters, Groningen), *Ab urbe condita* by Livy (libri XXI—XXX). From 1911 on: large historical water-colour paintings. After 1959: subjects pertaining to cultural history (published by Wolters, Groningen). References about his person and his work: in the newspaper *Trouw,* June 17, 1961, and the weeklies *Spiegel,* Febr. 19, 1961, and *Panorama,* Sept. 2, 1961.

K. 11 Three designs for illustrations for the book „Schetsen en verhalen uit Indiës verleden" by J. W. de Jongh en J. J. Moerman. (According to Isings these were never published.)

K. 11, 1 *Baljee, als buitengewoon lid van den Raad van Indië in de vergulden koets* . . . . (Baljee riding in the gilt coach as an extraordinary member of the East Indies Council . . . .).
Water-colours. Signed Isings. 18 × 25 cm.

K. 11, 2 *Een kijkje op 't verdek van De Houtman's schip bij het naderen van den hoek van Bantam* (A view of the deck of De Houtman's ship approaching the Corner of Bantam).
Water-colours. Signed Isings '16. 18 × 25,5 cm.

K. 11, 3 *In Prauwen gezeten, werden de Heeren (nl. de G.G. Maetsuyker en den Commandeur Cuneus) met hun gevolg naar de Peerl gevoerd* . . . (Seated in proa's the gentlemen (Governor General Maetsuyker and Commander Cuneus) were conveyed to De Peerl together with their attendants . . .).
Water-colours. Signed Isings. 18 × 25,5 cm.

Provenance: Presented by Wolters' Publishing Company, April 5, 1921.

KRAINZ, dr. C. F. S.

Born 1849. Febr. 18, 1875 appointed army surgeon in Atjeh. From 1878—1879 in Banda. 1879—1880 in Benkulen. 1880—1883 in Kediri. 1883—1886 in Pasuruhan.

O. 7 Sketch-book (15,5 × 23 cm) containing the following leaves.

O. 7, 1 *Santa Maria. Western Island.*
Pencil. Signed Dr. Krainz. Dated 16.7.1873.

O. 7, 2 *The Needles.*
Pencil. Signed Dr. Krainz. Dated 11/9 873.
Note: On the left hand side 'Wight', on the right hand side 'Lighthouse'.

O. 7, 3 *Stromboli.*
Pencil. Dated 25.3.875.

O. 7, 4 *Ophir.*
Pencil. Dated Fort de Kock 23.2.875.
Note: With caption 'Sumatra 0° 10′ N.B. 100° oost L.'

O. 7, 5 *Ausläufer des Bonigan-gebirges (Quellen des fluszes Sudicat?) gesehen von meinem Hause zu Bandar* (View of spurs of the Bonigan

Mountains (Source of the river Sudicat?) from my house in Bandar). Pencil. Signed Krainz. Dated 11 April 1878.
Note: With caption: a. Pasanggrahan, b. Padang Tinggie.

O. 7, 6 *Benteng with Dutch flag.*
Pencil. 10,5 × 12 cm.
Note: Possibly the military post at Bandar, which was stationed there from 1866—1906.

Provenance: Bought 1960.

LACOMBLÉ, L.

Born in Delft. Painter and etcher, perhaps son of the sculptor Eugène Lacomblé. He worked in Holland and Sweden.

O. 8, 1 *Sumbanese weaving-paterns, human motifs.*
Water-colours. Signed L. 22,5 × 16 cm.
Note: On reverse side: L. Lacomblé, 1904.

O. 8, 2 *Sumbanese weaving-patterns heron motifs.*
Water-colours. Signed L. 22,5 × 16 cm.
Note: On reverse side: L. Lacomblé, 1904.

LEBEAU, Joris Johannes Christiaan

Born May 26, 1878, Amsterdam. Died March-April, 1945, Dachau. Versatile artist; student of the Academy in Amsterdam; worked in Heemstede, Laren, and The Hague; visited the East Indies.

S. 18 *Soewardi.*
Orig. woodcut in black frame. Signed Chris Lebeau. Dated '23 Mei 1919'. 48 × 36,5 cm.
Note: Portrait of Ki Hadjar Dewantoro (formerly R. M. Suwardi Surjaningrat, 1889—1959), Indonesian politician and founder of the Taman Siswa schools, who was in exile in the Netherlands from 1913—1919. Print and frame of this woodcut were made by the artist himself.

Provenance: Bought from Mrs. M. S. Lebeau-Herman, 1966.

MARONIER, J. H.

Born Febr. 10, 1883, 's-Hertogenbosch.
Auto-didact. Professor A. Allebé, then director of the National Academy of Arts at Amsterdam, adjudicated his work and advised him to specialize in portrait-drawing and gave him some commissions. Later on Professor Vogelsang gave him his valued advice.

F. 9, 1 *M. W. F. Treub.*
Chalk. Signed JHM. Dated 1920. 36,5 × 27,5 cm.
Note: Minister of Finance, later chairman of the Board of Netherlands East-Indies Entrepreneurs.

F. 9, 2 *Prof. Ph. S. van Ronkel.*
Pencil. Signed JHM. Dated 1944. 37 × 18 cm.
Note: Professor at Leiden University, at that time chairman of the Royal Institute. Published in *B.K.I.* CX (1954), facing p. 285.

F. 9, 3 [*C. W. Th.*] *Baron van Boetselaer van Asperen tot Dubbeldam.*
Pencil. Signed JHM. Dated 1946. 27 × 22 cm.
Note: Mission director, committee member of the Royal Institute.

F. 9, 4 *Dr.* [*W. H.*] *Rassers.*
Pencil. Signed JHM. 32 × 25 cm.
Note: Director of the National Museum of Ethnology in Leiden, committee member of the Royal Institute. Drawn approx. 1946.

F. 9, 5 *Ir.* [*H.*] *Maclaine Pont.*
Pencil. Signed JHM. 35 × 26,5 cm.
Note: Architect of the T.H.S. (Technical College) in Bandung; secretary of the 'Vereniging Modjopahit' (Mojopahit Society). Drawn approx. 1948.

F. 9, 6 *Tampoebolon.*
Chalk. Signed JHM. 32 × 24,5 cm.
Note: Rajah from the Batak countries. Drawn approx. 1943.

F. 9, 7 *Javanese girl.*
Chalk. Signed JHM. 43 × 32 cm.

F. 9, 8 *Dr. F. W. Stapel. AET. 65.*
Pencil. Signed JHM. Dated 1944. 35 × 26,5 cm.
Note: Secretary and later chairman of the Royal Institute. Published *B.K.I.* 113 (1957), facing p. 113.

F. 9, 9 *Dr. P. Voorhoeve (profile).*
Pencil.. Signed JHM. Dated 1964. 40,5 × 33 cm.
Note: Secretary of the Royal Institute.

F. 9, 10 *Dr. P. Voorhoeve (front view).*
Pencil. 40,5 × 32 cm.
Note: Drawn approx. 1964.

K. 12, 1 *Negro, West-Indies.*
Pencil and red chalk. Signed JHM. 25,5 × 15,5 cm.

O. 9, 1—26 *Arms of the towns of the Netherlands East Indies.*
Water-colour paintings (mounted on cardboard, in port-folio).
11 × 14,5 cm.

Note: Arms of: Amboina, Bandung, Banjuwangi, Batavia, Blitar, Buitenzorg, Cheribon, Garut, Madiun, Magelang, Makassar, Malang, Manado, Medan, Modjokerto, Padang, Palembang, Pasuruan, Pekalongan, Salatiga, Semarang, Sukabumi, Surabaya, Tegal, Tjiandjur, West-Java. Copied from the archives of the Hoge Raad van Adel (High Council of Nobility).

MECHELEN, H. L. Ch. te

Born Oct. 5, 1841, Rembang. Died May 30, 1917, Bandung.
Civil service officer. Important authority on the Javanese language, and big-game hunter. From 1882—1885 assistant resident at Djuana, specially charged with subduing the opium-smuggling trade. May 1885 resident, temporarily charged with the supervision of the precautions at sea to put an end to such smuggling (Residèn Laut = Sea Resident), stationed in Semarang, later in Rembang. In 1889 chief inspector in charge of opium affairs. Went on leave in 1893 and in 1898 he received honourable discharge from service.

H. 423 *Kris uit de afdeling Pati, district Glanggong, gevonden in de desa Beketel in het jaar ......, was in het bezit van Mas Kromo Astro, als poesaka van zijn vader geërfd* (Kris from the sub-district of Pati, district of Glanggung, found in the village of Beketel in ......, was the possession of Mas Kromo Astro, who inherited it from his father as a sacred heirloom).
Water-colours. 34 × 21 cm.
Note: Approx. 1885.

Provenance: From the manuscript H. 423 of the Royal Institute.

References: "In memoriam" by G. P. Rouffaer, published in: *Indië Oud en Nieuw,* II, 1917—1918, p. 305—312.

MEETEREN BROUWER, Menno van

Born Nov. 1882, Zwolle.
Attended the National Academy of Arts 1904—1907. In 1910 he went to the Netherlands East Indies. Made political cartoons for newspapers and during his travels through Java, Sumatra and Bali he painted pictures. Published a book of pictures *Nederlandsch Indië zoals het was, Menno's Indische penkrabbels* (Dutch East Indies in former times, Menno's ink sketches of the Indies), Rijswijk 1956.

K. 13, 1 *Domestic scene in Java.*
Ink. Signed Menno. 32 × 25 cm.

Provenance: Bought at Vogel's antiquary, 1962.

MÜLLER, Georg

Born July 8, 1778, Mentz. Died Jan. 20, 1826, Borneo.
He joined Napoleon's Russian campaign as an officer. Entered the service of the Netherlands East Indies Army in 1816. Took part in the expedition against Sambas. Was appointed resident of the Residency of West Borneo. On an expedition along the Kapuas river he was murdered by Dayaks. Manuscript H. 34 with various notes by H. T. F. K. E. W. A. C. von Dewall contains 4 pencil drawings by G. Müller.

H. 34, Va *Gezigt op de bergen, waaruit de Kapoeas ontspringt. De houten beelden zijn terminus, Battok genaamd, opgerigt door de Dj. Kayan Sibau. Mond van de riv. Sibau* (View of the mountains where the Kapuas river rises. The wooden statues are called Battok (terminus), and have been erected by the Dayak Kayan Sibau. Mouth of the Sibau river).
Pencil. Dated 13 April 1824. 9 × 14 cm.
Note: Reproduced in: *De Indische Bij,* Vol. I, facing p. 353. Battok = probably Jav. *patok* = boundary mark.

H. 34, Vb *G. Müller's reis op de Kapoeas 13 April 1824. Ontmoeting bij de rivier Sibau en Monnai; Dayaks van Kayan, onder aanvoering van Tokoan broeder van Segau, hoofd der Daj. Sibau, Poenan, Kettan en Bokattan* (G. Müller's journey on the Kapuas, April 13, 1824. Meeting at the rivers Sibau and Monnai; Dayaks from Kayan led by Tokoan, brother of Segau, chief of the Sibau, Punan, Kettan and Bokattan Dayaks).
Pencil. 9 × 15 cm.
Note: River with boat, Dutch flag and proa. Reproduced in: *De Indische Bij,* Vol. I, facing p. 176.

H. 34, Vc *2 oudheden te Sangouw* (Two antiquities at Sanggau).
Pencil. Dated 12 Febr. 1824. 5 × 10 cm. and 8 × 8 cm.
Note: With note by Müller "Ook nog een olifantskop voorhanden en anderen" (Also present an elephant's head and other objects). Reproduced in: *De Indische Bij,* Vol. I, facing p. 196.

H. 34, Vd *Inscriptie op eene rots bij Sangouw. Oever der rivier Segayan al. Sangouw* (Inscription on a rock near Sanggau, bank of the Segayan or Sanggau river).
Ink. 6 × 15 cm.
Note: Reproduced in: *De Indische Bij,* Vol. I, facing p. 194.

Provenance: J. Hageman Jcz.

References: F. H. van Naerssen, De Batoe Sampai bij Sanggau West Afd. Borneo, *Cultureel Indië,* Vol. 3, July Aug. 1941.
H. Kern, *Verzamelde geschriften,* Vol. VII, p. 159.
C. W. Blume, Toelichtingen aangaande nasporingen op Borneo van G. Müller (Notes on G. Müller's research in Borneo), *De Indische Bij,* Vol. I, 1843, p. 103—176.
Ernst Müller, Levensbericht van Georg Müller (Obituary for Georg Müller), ibidem p. 177—196.
List of sketches made by G. Müller from 27 Jan.—13 April 1824, ibidem p. 154—160.
Georg Müller, Proeve eener geschiedenis van een gedeelte der Westkust van het eiland Borneo (First attempt at a historical description of part of the West Coast of Borneo), ibidem p. 197—375.

PALMER VAN DEN BROEK, Ch. L. J.

Born 1862. Died Sept., 1937.
Civil service officer. Went to the Dutch East Indies approx. 1888. In 1889 'aspirant-controleur' at Pajakumbo. In 1890 'controleur' 2nd class at Solok, in 1891 on the isle of Nias, 1894 in Kota Radja, 1895 at Segli, where he was appointed 'controleur' first class in the following year. 1900 at Muar Labuh, 1907 assistant resident at Fak-Fak, 1908 at Simalungun, 1912—1915 resident of Ternate. With his reports he often enclosed topographical sketches and drawings.

K. 14, 1 *Si Gala-Gala.*
Pencil. Dated March 25, 1888. 11,5 × 16,5 cm.

K. 14, 2 *Ruins of the Fort Pulu Tjinko. Coast of Painan.*
Pencil. 17,5 × 14 cm.

K. 14, 3 *Bay of Sibolga (?).*
Pencil. 13,5 × 24,5 cm.

K. 14, 4 *Large house at Hutarimbaru.*
Pencil. 19 × 25,5 cm.
Note: The numbers K. 14, 1—4 are from manuscript H. 592.

K. 14, 5 *Kilangan Hindi.*
Pencil. 15 × 20 cm.
Note: The drawing shows a small cane-press

K. 14, 6 *Zwaardgreep van Zuid-Nias* (Sword-hilt from South Nias).
Pencil. 10 × 10 cm.

K. 14, 7 *Speer* (Spear).
Pencil. 9 × 4 cm.

K. 14, 8 *Bewerking van een tôri* (Decoration on a tôri).
Pencil. 10 × 10 cm.
Note: The numbers K. 14, 5—8 are on the same sheet, and have explanatory texts.

Provenance: Presented by A. Palmer van den Broek.

PERS, Auguste van

Approx. 1850.
In June 1840 he was a draughtsman at the geographical department of the Algemene Secretarie (Secretary General's office) at Buitenzorg. From 1845 till 1847 he was a member of the staff of Dr. C. W. Schwaner during his activities in Borneo. Later on he worked in Java. The book *Nederlandsch Oost-Indische typen* (Dutch East-Indies types), 's-Gravenhage 1856, contains 44 lithographs copied from the drawings by A. van Pers. The first instalment of this book was published in 1853. Probably Van Pers had gone back to Holland by then. The source of these drawings made in Borneo, is the same as of those made by H. von

Gaffron He too probably contributed to the illustrations of Schwaner's *Borneo*. The portrait of "Tommongong van Poeloe-pétak (Nicodemus)" is probably his work. (Title page, Vol. I).

K. 15, 1 *Balia, oud 16 jaar, na eene 7 Jaarige Slavernij onder de Zeerovers Kort na haare bevrijding te Banjermassing den 18 November* (Balia, 16 years old, after 7 years' slavery under pirates, shortly after her release in Banjarmasin on November 18, [1845 or 1846]).
Pencil. Signed A. van Pers. 32 × 24,5 cm.
Note: Co-signature Dr. C. W. Schwaner, no. 1.

K. 15, 2 *Van de Kapoeas Rivier Borneo. Reis van Dr. Schwaner in de Maand Junij 1845* (From the Kapuas River, Borneo. Journey of Dr. Schwaner in the month of June 1845).
Water-colours. Signed A. van Pers. 30 × 23,5 cm.
Note: The drawing shows a rail babbler (Eupetes macrocerus borneensis), see B. E. Smythies, *The birds of Borneo,* picture XXXIII. Co-signature Dr. C. W. Schwaner, no. 2.

K. 15, 3 *Van de Kapoeas rivier Borneo. Reis van de Dr. Schwaner in de Maan Junij 1845* (From the Kapuas River, Borneo. Journey of Dr. Schwaner in the month of June 1845).
Water-colours and pencil. Signed A. van Pers. 27,5 × 22 cm.
Note: The drawing shows a baldheaded wood shrike (Pityriasis gymnocephala), see B. E. Smythies, *The birds of Borneo,* picture XLV. Co-signature Dr. C. W. Schwaner, no. 3.

F. 10, 1 *Banjermassing. Gezigt van de Boom (tolhuis) op de Rivier van Banjermassing. Reizen van Dr. Schwaner op Borneo* (Banjarmasin. View of the toll-bar on the river of Banjarmasin. Dr. Schwaner's journeys in Borneo). Pencil on greyish paper. Signed A. Van Pers ad nat. delineavit. Dated 'Maand December 1846'.
Note: Co-signature Dr. C. W. Schwaner, no. 5.

K. 15, 4 *Wealthy Buginese.*
Pencil and water-colours. Signed A. van Pers fecit. 26 × 21 cm.
Note: The original caption has probably been cut off and has been replaced by a pencil note on the reverse side. This note mentions the co-signature Dr. C. W. Schwaner, no. 10.

K. 15, 5 *Captain of the Chinese (from Banjarmasin?).*
Water-colours on greyish paper. Dated (on reverse side) approx. 1844. 33 × 21 cm.
Note: Pasted on the drawing a co-signature Dr. C. W. Schwaner, no. 12. The original caption has probably been cut off and has been replaced by a pencil note on the reverse side.

K. 15, 6 *Kampong (village) on the river, Banjermasin.*
Pencil on brown paper. Signed A. van Pers. Dated 1847. 17 × 30,5 cm.

K. 15, 7 *Phylloruis cyanopogon boie n. sp.*
Water-colours and ink. 22,5 × 20,5 cm.

Note: The drawing shows a bird study. On the reverse side is a note: 'teekening vermoedelijk door A. van Pers' (drawing probably by A. van Pers).

F. 10, 2 *View of a river in Borneo.*
Pencil and water-colours. 23 × 38,5 cm.
Note: On reverse side note: vermoedelijk door A. van Pers (probably by A. van Pers).

K. 15, 8 *Banjarese prince.*
Pencil. 26 × 21 cm.
Note: Sultan Adam, reigned 1842—1857 (?). On reverse side is a note: teekening vermoedelijk door A. van Pers (drawing probably by A. van Pers) ca. 1844.

F. 10, 3 *Title page of the book "Nederlandsch Indische Typen"; geteekend door A. van Pers. Batavia, 1851.*
Ink. Signed A. van Pers sculpsit, St. dr. M. Degens. 37,5 × 28 cm.

PFAFF, L. (See C. B. H. von Rosenberg, K. 18, 34.)

K. 16, 1—22 *Verzameling van Mohammedaansche oudheden die gevonden worden in en om Grisseé op Java. Naar Javaansche origineele stukken getrouw geteekend door A. van Pers en geannoteerd door den Wel Ed. Heer Hageman* (Collection of Mohammedan antiquities found in and around Gresik in Java. Accurately copied from original Javanese objects by A. van Pers, with notes by (J.) Hageman (J. Cz.)).
Ink on Japanese paper. 34 × 41 cm.
Note: 22 drawings bound. With Dutch and Javanese captions. Probably based on original Javanese drawings about 1850, for inclusion in a publication which never appeared.

K. 16, 1—16 *Graven* (Graves).

K. 16, 1 *De lange graven te Lerén (1399)* (The long graves at Leran (1399)).

K. 16, 2 *Malik Ibrahim (1419).*
Note: Grave at Gresik.

K. 16, 3 *Van Rodjob-pendito te Woenoet (1463)* (Rodjo-pendito at Wunut (1463)).

K. 16, 4 *Niai-gédé (Voorname vrouw) Penatée (1480) kamp. Keboengsoem* (Niai-gédé (distinguished lady) Penatih (1480) kampung Kebungson).

K. 16, 5 *Van Soenan Ratoe Giri (1518)* (Sunan Ratu Giri (1518)).

K. 16, 6 *Njai-gédé (Voorname vrouw) Koekoesan (1599)* (Njai-gédé (distinguished lady) Kukusan (1599)).

K. 16, 7 *Pangéran Babeloek (1600). Kamp. Karangkireng* (Pangéran Babeluk (1600). Kampung Karangkering).

K. 16, 8 *Soesoehoenan Prapén (1626)* (Susushunan Prapèn (1626)).

K. 16, 9 *Kiai-Tommongong (voorname man) Poesponegoro (1721). Kamp. Gapoerob* (Kiai-Tommongong (distinguished gentleman) Pusponegoro (1721). Kampung Gapurob).

K. 16, 10 *Van Kiai-Tommongong (voorname man) Soeronegoro (1768). Van Kiai-Tommongong (voorname man) Kertoredjo (1772). Van Kiai-adipati (voorname vrouw) Hardjonegoro (1810)* Kiai-Tommongong (distinguished gentleman) Suronegoro (1768). Kiai-Tommongong (distinguished gentleman) Kertoredjo (1772). Kiai-adipati (distinguished lady) Hardjonegoro (1810)).

K. 16, 11 *Van Kiai-gédé (voorname man) Harenareng. Kamp. Sedho paksoei* (Kiai-gédé (distinguished gentleman) Harenareng. Kampung Sido pakso).

K. 16, 12 *Van Kiai-adipati (voorname man) Brotonegoro, Grissée* (Kiai-adipati (distinguished gentleman) Brotonegoro, Gresik).

K. 16, 13 *Njai-gédé (voorname vrouw) Pamengkang. Kamp. Soemoersingo* (Njai-gédé (distinguished lady) Pamengkang. Kampung Sumursingo).

K. 16, 14 *Pangéran Hangoenbayo berg Passabilan* (Pangéran Hangunbayo mountain Pasabilan).

K. 16, 15 *Njai-gédé (voorname vrouw) Toempang* (Njai-gédé (distinguished lady) Tumpang).

K. 16, 16 *Njai-gédé (voorname vrouw) Goenongandjer* (Njai-gédé (distinguished lady) Gunongandjar).
Note: Gunung-anjar?

K. 16, 17—19 *Tempels* (Temples).

K. 16, 17 *De Tempel te Grissée (1570)* (The temple at Gresik (1570)).

K. 16, 18 *De tempel in het dorp Goemenol* (The temple in the village of Gumenol).
Note: Gumen?

K. 16, 19 *De Tempel op den heuvel Giri-gadja* (The temple on Giri-gadjah hill).

K. 16, 20—22 *Stukken* (Objects).

K. 16, 20 *De heilige dolk Soeroe Angoonangan (1583)* (The holy dagger Sura Angun-angun (1583)).

K. 16, 21 *De steen voor het graf van Soeronegoro (1768) in de wijk Gapoero* (Tombstone at the grave of Suronegoro (1768) in the quarter of Gapura).

K. 16, 22 *De steen voor het graf van Poesponegoro [1721] in de wijk Gapoero* (Tombstone at the grave of Pusponegoro [1721] in the quarter of Gapura).

Provenance: Bought at Van Stockum's antiquary, 1913.

RAPPARD, Jhr. Josias Cornelis

Born April 24, 1824, Nijmegen. Died May 17, 1898, Leiden.
Arrived in the Netherlands East-Indies in 1842 as a 2nd lieut. inf. Later on he became a colonel, and stayed in the Indies till 1872. He made many studies in the Indies on which later on he based his water-colour paintings as, for example, the ones which were reproduced as lithographs in the book *Nederlandsch Indië* (Netherlands Indies) by W. A. van Rees, Leiden 1881.
F. 14, 14 *Reception at the Palace of Buitenzorg.*
Water-colours. 17 × 23 cm.
Note: Or reception at Concordia Club in Batavia? Not included in *Nederlandsch Indië.*

Provenance: Bought in 1962.

RIEKERK, G. H. M.

'Controleur' in the civil service.

K. 17, 1 *Klemmen voor het uitsparen van versiering op te verven mangitjobladeren, oorversiersel, Soemba* (Clamps which are placed on mangitjo-leaves in order to leave a pattern after dyeing, ear ornament, Sumba).
Water-colours. Light brown tinted paper. 24,5 × 40 cm.
Reference: G. H. M. Riekerk, "Haboe Hoegho", *B.K.I.* 99 (1940) p. 89—99.

ROSENBERG, Carl Benjamin Hermann von

Born April 7, 1817, Darmstadt. Died Nov. 15, 1888, The Hague.
Served in the Dutch East Indies army from 1840 to 1871, the first sixteen years in Sumatra, and made many journeys to explore (parts of Sumatra and the eastern part of the archipelago), partly by order of the Government. His collected reports which had appeared individually in Dutch were afterwards published in German in his book *Der Malayische Archipel; Land und Leute in Schilderungen, gesammelt während eines dreiszigjährigen Aufenthaltes in den Kolonien* (The Malay Archipelago; pictures of the land and its people collected in the course of a thirty years' stay in the colonies), Leipzig 1878.

K. 18 *Origineele teekeningen, behoorende bij het werk "Der Malayische Archipel" door H. von Rosenberg* (Original drawings of illustrations for the book "Der Malayische Archipel" by H. von Rosenberg).

These drawings were all but one reproducel as wood-engravings in his book (below referred to as M.A.) and some of them reproduced again as lithographs, in "Nieuw Guinea, ethnographisch en natuurkundig onderzocht en beschreven in 1858 door een Nederlandsch Indische commissie" (New Guinea, ethnographic and physical studies and descriptions made in 1858 by a Netherlands East-Indies Commission), published in 1862 as vol. IX of *B.K.I.* and below referred to as *Nieuw Guinea.* No. K. 18, 33 has been reproduced in this book only. Nos. K. 18, 34—38 have not been drawn from nature but are copies based on photographs. As No. K. 18, 38 is signed 'L. Pfaff gez. 1875', and therefore not by Rosenberg, it is possible that the other four are not by Rosenberg either.

Sumatra

K. 18, 1 *Hangende brug over de Batang Torro (Angola)* (Suspension bridge across the Batang Toru (Angkolla)).
Pencil. 16 × 23 cm.
Note: M.A., p. 19: Hängebrücke in Padang-lawas.

K. 18, 2 *Bataksche tooverstok* (Batak magic staff).
Pencil. 27 × 9 cm.
Note: M.A., p. 41: Kriegspanier der Battas.

K. 18, 3 *Dorp Lobangtunkung aan den Simpang Sinoendang* (Village of Lobang-tunkung on the Simpang Sinundang).
Water-colours. 11 × 18 cm.
Note: M.A., p. 49: Dorf Lobangtunkung in der Landschaft Simpang-Sinundang.

K. 18, 4 *Battak van Dairi ter markt gaande* (Batak from Dairi on his way to market).
Water-colours. 19,5 × 15 cm.
Note: M.A., p. 56: Batta von Pak-Pak, Benzoë zu Markte bringend.

K. 18, 5 *Battak van Dairi in krijgsrusting* (Batak from Dairi in battle array).
Water-colours. 20 × 11 cm.
Note: M.A., p. 57: Batta von Pak-Pak, mit Lanze und Schild bewaffnet. Letzterer befindet sich im Darmstädter Museum.

K. 18, 6 *Battaks tooverinstrument* (Batak magic instrument).
Pencil. 23 × 13,5 cm.
Note: M.A., p. 59: Batta'sches Zauberinstrument aus Büffelhorn. Im Darmstädter Museum.

K. 18, 7 *Graf van een Battakhoofd in Dairi* (Grave of a Batak chief in Dairi).
Water-colours. 14 × 10 cm.
Note: M.A., p. 61: Battagrab in der Nähe von Lubang-tungkung.

The islands to the West of Sumatra

K. 18, 8 *Nias. Vrouw uit het volk* (Village woman from Nias).
Water-colours. 17 × 10,5 cm.
Note: M.A., p. 145: Frau aus dem Volke von Nord-Nias.

K. 18, 9 *Nias. Vrouw van een hoofd* (Nias. Wife of a chief).
Water-colours. 17 × 10,5 cm.
Note: M.A., p. 159: Frau eines Häuptlings in Festkleidung.

K. 18, 10 *Huis van Noord-Nias* (House in North Nias).
Water-colours. 17,5 × 14 cm.
Note: M.A., p. 151.

K. 18, 11 *Graf op Zuid-Nias* (Grave in South-Nias).
Water-colours. 16,5 × 12 cm.
Note: M.A., p. 157: Leichengestell mit Sarg bei Orahili auf Süd-Nias.

K. 18, 12 *Hoofd van Zuid-Nias* (Chief from South-Nias).
Water-colours. 17 × 10,5 cm.
Note: M.A., p. 161: Häuptling von Sendegeassi in Festkleidung.

K. 18, 13 *Verschillende kano's van Mentawei* (Various canoes from Mentawei).
Lithograph (of a water-colour painting?). 12 × 19 cm.
Note: M.A., p. 177.

K. 18, 14 *Man in krijgsrusting van Mentawei* (Man from Mentawei in battle array).
Water-colours. 19 × 12 cm.
Note: M.A., p. 185: Mann von Siberut zum Streite gerüstet.

K. 18, 15 *Vrouw van Mentawei* (Woman from Mentawei).
Water-colours. 18 × 10 cm.
Note: M.A., p. 185: Frau von den Pageh-Inseln.

K. 18, 16 *Mannen van Mentawei* (Men from Mentawei).
Phototype or lithograph. 11,5 × 15,5 cm.
Note: M.A., pp. 192 and 185: Mann von Pora und Mann von den Pageh-Inseln.

K. 18, 17 *Dorp op Mentawei* (Village in Mentawei).
Pencil. 10,5 × 22 cm.
Note: M.A., p. 200: Das Dorf Silal auf Nord-Pâgeh.

K. 18, 18 *Tuinhuis op Mentawei* (Hut in ricefields in Mentawei).
Pencil. 18 × 10 cm.
Note: M.A., p. 201: Gartenhaus bei Sibiribenua.

K. 18, 19 *Man en vrouw van Engano* (Man and woman from Engano).
Water-colours. 18 × 12,5 cm.
Note: M.A., p. 208.

K. 18, 20—21 *Vrouwenkapsel van het eiland Engano* (Woman's head-dress from Engano).
Pencil. Both 12,5 × 11 cm.
Note: M.A., p. 209: Kopfzierrate der Frauen auf Engano. Originale im Darmstädter Museum.

Celebes

K. 18, 22 *Meer van Linoe (Menado). Plaats waar graaf Vidua [de Conzana in der 40-er jaren der 19e eeuw] verongelukte* (Lake of Linu (Menado). Count Vidua [de Conzana] met his death in an accident here [in the nineteen forties]).
Pencil. 7,5 × 10,5 cm.
Note: M.A., p. 265: Schlammsee von Linu . . .

The Molucca Archipelago

K. 18, 23 *Landverkenningen op de zuidkust van Seram* (Landfalls on the south coast of Seram).
Pencil. Both 7,5 × 12 cm.
Note: M.A., p. 296: Berg Srewa bei Osson mit dem Ukele im Hintergrund. Berg Kissi bei Batuassa. Two drawings on one sheet.

K. 18, 24 *Landverkenningen op de zuidkust van Seram* (Landfalls on the south coast of Seram).
Pencil. 16 × 11 cm.
Note: M.A., p. 297: Berg Téri von Westen gesehen. Berg Téri von Osten gesehen. Two drawings on one sheet.

K. 18, 25 *Gezicht in de bocht van Sawaai* (View of the bay of Sawai).
Water-colours. 7 × 21 cm.
Note: M.A., p. 301: Ansicht des Berges Subun, zur Rechten im Hintergrunde der Hatu-sakka, von Sleman aus gesehen.

K. 18, 26 *Ruïne van het oude Compagniesfort in de bocht van Sawaai* (Ruins of the fort of the East Indies Company in the bay of Sawai).
Pencil. 17,5 × 12,5 cm.
Note: M.A., p. 304.

K. 18, 27 *Alfoer van Seram* (Alfur from Seram).
Pencil. 17 × 11,5 cm.
Note: M.A., p. 313: Alfure von Marhunu.

New Guinea

K. 18, 28 *Doodengrot bij Kaimani (Z.W.k. Nieuw-Guinea)* (Burial cave near Kaimani (S.W. coast of New Guinea)).
Pencil. 12 × 18,5 cm.
Note: M.A., p. 417; *Nieuw Guinea,* D.

K. 18, 29 *Dorp Timbona (Etnabaai* (The village of Timbona ,Etna bay)). Phototype or lithograph. 11,5 × 19 cm.
Note: M.A., p. 424.

K. 18, 30 *Roemsram bij Doré* (Rumsram near Doré).
Pencil and water-colours. 9,5 × 17,5 cm.
Note: M.A., p. 456: Das Rumsram bei Rasambôri; *Nieuw Guinea*, S.

K. 18, 31 *Papoes graf bij Doré* (Papuan grave near Doré).
Water-colours. 14 × 9 cm.
Note: M.A., p. 457: Papu'sche Grabstätte bei dem Dorfe Rasambôri; *Nieuw Guinea*, PP (without the standing Papua).

K. 18, 32 *Papoea van Dorei* (Papua from Dorei).
Pencil. 17 × 15 cm.
Note: M.A., p. 473: Papua der Humboldtbai.

K. 18, 33 *Tempel in de Humboldbaai* (Temple in Humboldt Bay).
Water-colours. 15,5 × 10,5 cm.
Note: Nieuw Guinea, FF.

Drawings based on photographs from Gorontalo and the North Moluccas

K. 18, 34 *Portret van Monoarfa, Radja van Gorontalo* (Portrait of Monoarfa, Rajah of Gorontalo).
Pencil. 15,5 × 12 cm. Signed L. Pfaff gez. (= drawn) 1875.
Note: M.A., p. 225. Edges cut off.

K. 18, 35 *Jonkman van Gorontalo* (Young man from Gorontalo).
Pencil. 17,5 × 12,5 cm.
Note: M.A., p. 232: Junger mann aus dem besseren Stande.

K. 18, 36 *Danseres van Gorontalo* (Dancing-girl from Gorontalo).
Pencil. 20 × 15 cm.
Note: M.A., p. 233: Tänzerin des Radja von Gorontalo.

K. 18, 37 *Alfoeren van Halmahera* (Alfurs from Halmahera).
Pencil. 18 × 15 cm.
Note: M.A., p. 288: Alfurus von Elpaputi.

K. 18, 38 *Lijfwacht van den sultan van Ternate* (Body-guard of the Sultan of Ternate).
Pencil. 18,5 × 15 cm.
Note: M.A., p. 401.

Provenance: Presented by C. B. H. von Rosenberg.

RIJCK, A. G. van

'Tweede pakhuismeester' (deputy warehouse superintendant) in Batavia, 1859—1875.

K. 19 *Album containing caricatures, probably relating to the assistant resident J. B. Westenberg.*
Pencil. Drawn approx. 1867—1869. 33 × 21 cm.
Note: Several of the drawings contain detailed explanations.

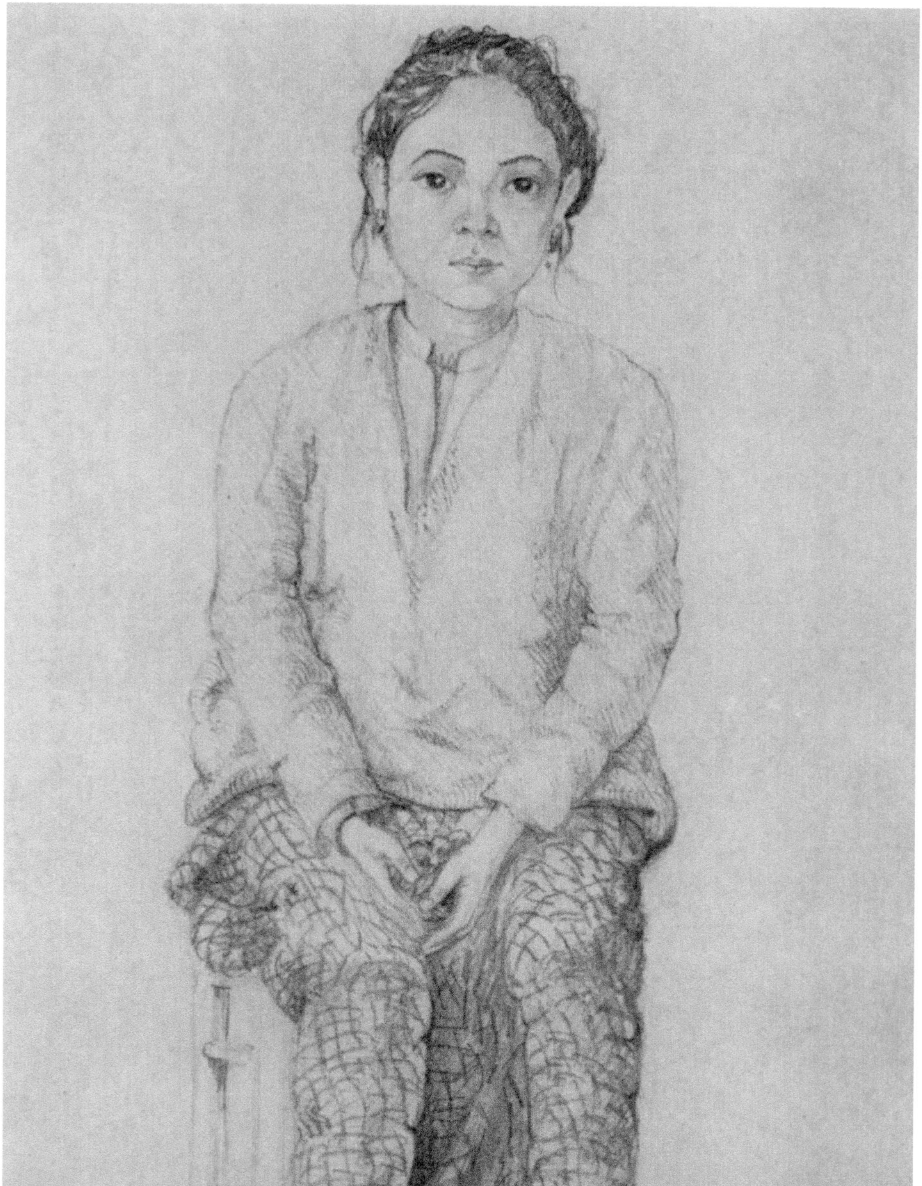

A. van Pers   Balia, after her release from slavery   K. 15, 1

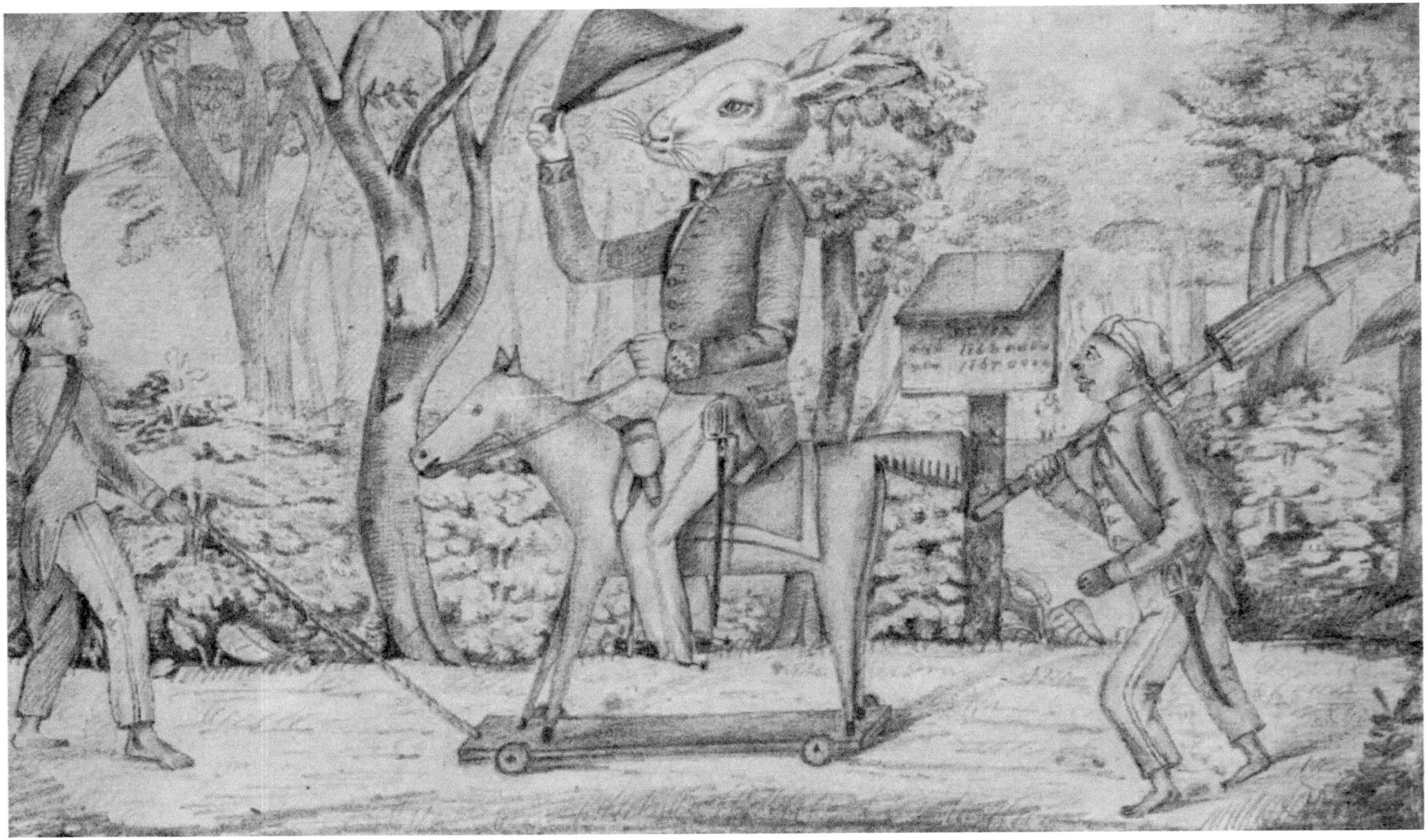

A. G. van Rijck The inspector of coffee plantations K. 19, 4

K. 19, 1 *Waarmêe ezels in Indië vooruitkomen!* (How asses make progress in the Indies!).
11 × 19,5 cm.

K. 19, 2 *Met het ambt komt het verstand!* (Practice makes perfect).
15,5 × 13,5 cm.

K. 19, 3 *Al draagt een aap een gouden ring, hij is en blijft een lelijk ding! (Jacob Cats)* (An ape 's an ape, a varlet 's a varlet, though they be clad in silk and scarlet).
20 × 17 cm.

K. 19, 4 *Hoe vreesachtige naturen koffijtuinen kunnen inspekteren* (How timid characters may safely inspect coffee plantations).
17 × 28 cm.

K. 19, 5 *Hoe Judas Barbaras Westendal in zijne afdeeling op tourné gaat!* (How Judas Barbaras Westendal goes on an official tour of his district!).
15,5 × 28 cm.

K. 19, 6 *Medelijden* (Compassion).
19,5 × 19,5 cm.

K. 19, 7 *De geschiedenis van twee Uilen* (The story of two Owls).
17 × 17 cm.

K. 19, 8 *Hoe vrijwillige stortingen werden verkregen* (The way to be given voluntary payments).
13 × 18 cm.

K. 19, 9 *De kikker en de stier* (The frog and the ox).
21 × 16 cm.

Provenance: Bought at the auction-sale of the house of Burgersdijk en Niermans and presented to the Institute by Mrs. J. de Loos-Haaxman in 1958.

SALM, Abraham

Born Oct. 29, 1801. Died Dec. 4, 1876.
Was a merchant in Surabaya. In Java he made drawings from nature and back in Holland he made oilpaintings based on these. He was a member of the Royal Academy of Arts in Amsterdam. He had great talent and an eye for the atmosphere of the tropics. Part of his work was reproduced in *De Indische Archipel* and *Java naar schilderijen en teekeningen door A. Salm,* Amsterdam, 1872, which is a collection of 24 prints of Javanese landscapes.

O. 10 *Sketch-book in marbled binding.*
18 × 23 cm.

O. 10, 1 *Bird.*
Water-colours.

O. 10, 2 *Landscape.*
Water-colours.
Note: Compare *Java,* print No. 9.

O. 10, 3 *Various proas from different districts among which a long travelling-proa.*
Water-colours and pencil.

O. 10, 4 *Tree in Indonesian landscape.*
Water-colours.

O. 10, 5 *Deer.*
Water-colours.

O. 10, 6 *Halting-place for mail-carts.*
Water-colours.

O. 10, 7—8 *Crocodiles.*
Water-colours and pencil.
Note: Unfinished.

O. 10, 9 *Old tree.*
Water-colours.

O. 10, 10 *Native houses in the country-side.*
Water-colours and pencil.

O. 10, 11 *Mountain landscape with native house and country-house.*
Water-colours and pencil.

O. 10, 12 *Landscape with houses.*
Water-colours and pencil.
Note: Cursory sketch.

O. 10, 13 *Sheep.*
Pencil.

O. 10, 14 *Landscape with river and bridge.*
Water-colours and pencil.

O. 10, 15 *Road along a ravine.*
Water-colours.
Note: Puntjak-pass?

O. 10, 16 *Waterfall.*
Water-colours.

O. 10, 17 *Buffalo-cart and buffalo.*
Pencil.

O. 10, 18 *Coffee-estate in West Java (?).*
Water-colours.

O. 10, 19 *Covered bridge near a rapid.*
Water-colours.

O. 10, 20 *Mountain village with high mountains in background.*
Water-colours.

O. 10, 21 *Landscape.*
Water-colours.
Note: Karbouwengat?

O. 10, 22 *View of the Tji-Liwung with two bridges.*
Water-colours.

O. 10, 23 *Landing-stage in a bay near Singapore (?).*
Water-colours and pencil.
Note: Drawing on a loose sheet.

Provenance: Bought at Loose's antiquary. The Hague, 1963.

SAYERS, Charles

Born 1901, Kendal (Java). Died 1943, Thambyăzajat (Siam).

Trained at the National Academy of Arts in Amsterdam. Was presented with the Cohen Gosschalk prize. He did his painting in Java, Bali and Sumatra.

S. 15 *Balinese dancer.*
Oil on canvas. Signed Ch. Sayers. 185 × 135 cm.

Provenance: Bought at exhibition of the 'Indische Kunstkring', The Hague, Jan. 1960.

References: Article by Tjalie Robinson in *Elseviers Weekblad,* Jan. 23, 1960.

SCHULTZE, [Paul?]

Approx. 1897.

K. 20, 1 *De Goenoengan-Lanang, een der spijsbergen, die op een Garĕbĕg-dag door 24 man wordt opgedragen en op last van den Sultan [van Jogjakarta] aan diens onderhoorigen ten geschenke gegeven* (The Gunungan-Lanang, one of the piles of food carried by 24 men on Garĕbĕg-day and given to the subordinates of the Sultan [of Jogjakarta] by his orders).
Lithograph. Signed Schultze. Dated [18]97. 23 × 31 cm.
Note: Numbered 5, chapter X. Presumably intended as an illustration for a book about Jogjakarta by Engelbert van Beevervoorde. The book was, however, never finished.

SCHWANER, Carl Anton Ludwig Maria

Born 1817, Mannheim. Died March 30, 1851, Batavia.
Geologist and explorer. Studied geology and mineralogy at the University

of Heidelberg (1836—41). Worked at the Museum of biology in Leyden. In 1841 appointed member of the Commission for Scientific Research. He arrived in the Indies in 1842 and was charged by the Government with the exploration of coal layers in South-East Borneo, but gradually he devoted himself more to the solution of ethnological and geographical problems with respect to Borneo in connection with the political tension between the Netherlands and the United Kingdom relating to this island. His greatest feat was the crossing of Borneo from Banjarmasin to Pontianak in 1847—48. Then he returned to Batavia to work out his scientific reports. He contributed to the foundation of the "Natuurkundige Vereeniging" (Scientific Association) and was its first president. After the completion of his reports he was charged with the resumption of his exploration of Borneo, but died when still in Batavia. He was assisted by the artists H. von Gaffron (1843—46) and A. van Pers.
In his book *Borneo, beschrijving van het stroomgebied van den Barito en reizen langs eenige voorname rivieren van het Zuidoostelijk gedeelte van dat eiland,* Amsterdam, 1853, he described in Vol. I: South-East Borneo on the basis of his own and other people's observations; in Vol. II he reported in diary form on his great, most famous journey from Banjarmasin to Pontianak.
There is no indication that the illustrations in the book relate to the great journey. It may be assumed that most of these were based on sketches by Schwaner's draughtsman H. von Gaffron.

K. 21, 1—10 *Trachten der Boeginesen von Pagattan* (National costumes of the Bugis from Pagattan).
10 pencil-drawings. Signed Schw. 34 × 21 cm.
Note: The first 3 are not accompanied either by captions or notes. The rest, numbered 5—11, have German captions and notes.

K. 21, 1 *Elderly woman.*
Dated Pagattan 8/9. 44.

K. 21, 2 *Younger woman.*
Dated Pagattan 9/9. 44.

K. 21, 3 *Woman and girl*
Dated Pagattan 8/9. 44.

K. 21, 4 *Kleidung des gegenwärtigen Imam von Pagattan* (Wearing apparel of the present Imam of Pagattan).
Dated Pagattan 1/9. 44.

K. 21, 5 *Ein Tjilabaij* (Priest in women's dress).
Dated Pagattan 9/8. 44.

K. 21, 6 *Kleidung der reicheren Boeginesen, wenn sie zum Gefechte gehen* (Battle attire of well-to-do Bugis).
Dated Pagattan 5/9. 44.

K. 21, 7 *Kleidung der Sklaven und der ärmeren Boeginesen* (Slaves' attire and wearing apparel of the poor Bugis).
Dated Pagattan 5/9. 44.

K. 21, 8 *Eine Frau aus dem unteren Stande des Volkes* (A woman of the lower classes).
Dated Pagattan 5/9. 44.

K. 21, 9 *Kleidung eines wohlhabenden Boeginesen* (Costume of a well-to-do Bugis).
Dated Pagattan 6/9. 44.

K. 21, 10 *Eine Frau aus der wohlhabenden Klasse* (A woman of the upper classes).
Dated Pagattan 6/9. 44.

Provenance: Presented by the Ministry of the Colonies, Aug. 11, 1858. La. A. no. 2 (See: Von Gaffron).

SLOET, A.

Probably Anton, baron Sloet van Oldruitenborgh, born 1823, died 1894. He is known to have been in Surabaya from 1849. He was the head of the sugar estate of Purwadadi near Madiun from 1853—1864.

F. 11, 1 *Missigiet te Soerakarta* (Mosque in Surakarta).
Pencil. Dated 1847. 22,5 × 27 cm.
Note: The Surambi is not shown in the picture.

F. 11, 2 *Chinesche graven te Tjebres (Soerakarta) nabij de plaats waar de troepen schijfschieten* (Chinese graves in Djebres (Surakarta), near the military shooting-range).
Pencil. Dated approx. 1847. 22 × 24,5 cm.

F. 11, 3 *Gezigt op de Merapi en Marbaboe van de vlakte van Djoeroek bij Solo* (View of the Merapi and Marbabu from the plain of Djuruk near Solo).
Pencil. Dated 1847. 18 × 27 cm.

F. 11, 4 *Gezigt op de Merapi van Toempak op de helling der Marbaboe gelegen* (View of the Merapi from the slope of the Marbabu in Tumpak).
Pencil. Dated 1847. 18 × 24,5 cm.

F. 11, 5 *Ingang naar de eerste terras der oudheden op Soekoeh* (Entrance to the first terrace of the ancient Sanctuary at Sukuh (Surakarta)).
Pencil. Dated ca. 1847. 18 × 27,5 cm.

F. 11, 6 *Oudheden op Tjetto* (Antiquities of Tjeta).
Pencil. Dated Solo 1847. 18 × 26,5 cm.
Note: With explanations.

F. 11, 7 *Ruine van een lustslot van den Sultan van Djocjacarta, Radjo Kesoemo genaamd* (Ruins of a pleasure-house of the sultan of Jogjakarta, named Radja Kesoema).
Pencil. Signed A. Sloet 1848. 18 × 28 cm.

F. 11, 8 *Zuidelijke poort der Kraton van Djocjakarta* (Southern gate of the Kraton (palace) of Jogjakarta).
Pencil. 17 × 25 cm.

F. 11, 9 *De witte paal te Djocja* (The (so-called) Witte Paal (White Pillar) in Jogjakarta).
Pencil. Dated 1848. 16,5 × 25 cm.

F. 11, 10 *Wilde pisang te Tandjong Tirto, Djocjacarta* (Wild banana tree in Tandjong-Tirto, Jogjakarta).
Pencil. 19 × 14 cm.

F. 11, 11 *Gezicht op de Aloen Aloen van de Kraton van den Soesoehoenan van Soerakarta* (View of the aloon-aloon of the Kraton (palace) of the Susuhunan of Surakarta).
Pencil. Dated ca. 1848. 25 × 37,5 cm.

Provenance: From an album of photographs of the Near East (No. 3—138).

STROYKENS

One pencil-drawing and 17 water-colours, several of which are signed "Stroykens", are to be found in the manuscript H. 117 "Een blik in de Residentie Palembang", dated Muntok, March 1, 1854; by L. Ullmann.

a *Lamatang* (Lematang).
Water-colours. 29 × 20 cm.
Note: The drawing shows two pile-dwellings.

b *Missigit (Ranas)* (Mosque (Ranas)).
Water-colours. 20 × 32 cm.

c *Goemei Oeloe* (Gumei Ulu).
Pencil. 20 × 32 cm.

d *Ogan.*
Water-colours. Signed Stroykens. 32 × 20 cm.
Note: 2 water-colour drawings on one sheet .The drawings show houses.

e *Moeara doea* (Muara Dua).
Water-colours. Signed Stroykens. 20 × 40 cm.
Note: The drawing shows a suspension bridge.

f *Bedok, Keto en Isiran* (Bedok, Keto and Isiran).
Water-colours. 20 × 32 cm.
Note: Two of the instruments shown are drums.

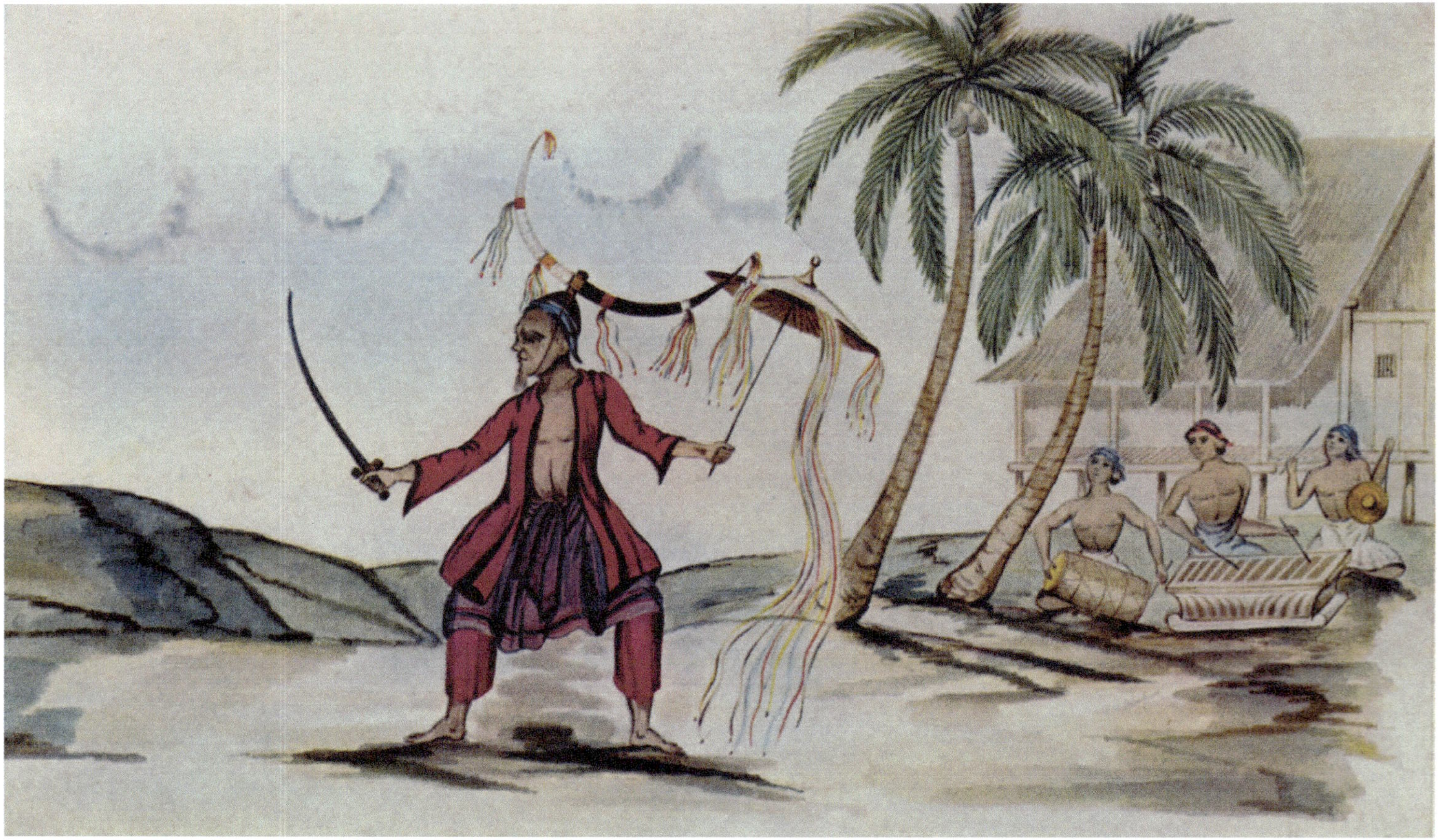

STROYKENS Panglima dalam, Ranas H. 117, q

g *Palembang.*
Water-colours. Signed Stroykens. 20 × 36 cm.
Note: The drawing shows the residency near the river.

h *Klingie.*
Water-colours. Signed Stroykens. 20 × 32 cm.
Note: The drawing shows a fort.

i *Moeara Roepit* (Muara Rupit).
Water-colours. 20 × 32 cm.
Note: The drawing shows a village on the river.

j *Tebing Tingi.*
Water-colours. Signed Stroykens. 20 × 32 cm.

k *Lahat.*
Water-colours. Signed Stroykens. 20 × 32 cm.

l *Kentjil Ayer* (Water-wheel).
Water-colours. 26 × 20 cm.

m *Kommering Oeloe* (Komering Ulu).
Water-colours. Signed Stroykens. 20 × 36 cm.
Note: The drawing shows a cock-fight.

n *Wisselaar* (Money-changer).
Water-colours. Signed Stroykens. 19 × 26,5 cm.

o *Gadis* (Girls).
Water-colours. 32 × 20 cm.
Note: The drawing shows a dance.

p *Kommering Oeloe* (Komering Ulu).
Water-colours. 20 × 32 cm.
Note: The drawing shows a scene with three persons and a house in background.

q *Panglima dalam (Ranas).*
Water-colours. 19 × 32 cm.
Note: The drawing shows dancing with a gamelan orchestra.

r *Lankayap. Tjoeroep ambinan.*
Water-colours. Signed Stroykens. 20 × 32 cm.

Provenance: Presented by A. L. de Sturler, Leiden, Febr. 15, 1896.

SUHR, C.

Painter in Hamburg approx. 1813; one of two brothers, Christoph and Cornelius, both painters in Hamburg. The former was born May 29 or 31, 1771, and died May 13, 1842; the latter was born January 8, 1781, and died July 3, 1857. There are few paintings by Cornelius known to exist, and therefore it is most likely that Christoph was the painter of Van Hogendorp's portrait.

S. 12 *General Dirk van Hogendorp.*
Oil on canvas. Signed: 'paint par C. Suhr'. To be dated about 1813/14. 240 × 140 cm.

Description: Dirk van Hogendorp (1761—1822) in a general's uniform, full-length, his right hand resting on a ground-plan of the fortifications of Hamburg of which town he was governor from 1813—1814. In the background to the left a bust of Napoleon I.

Provenance: Presented by Jkvr. M. A. van Hogendorp to the Indisch Genootschap, later transferred to the Royal Institute.

References: J. A. Sillem, *Dirk van Hogendorp (1761—1822) . . .* Amsterdam, 1890. On page 317—318 this portrait is mentioned to be hanging in the house in Brazil, see S. 13.
S. Kalff, "Herinneringen aan D. van Hogendorp", *Ned. Indië Oud en Nieuw,* Vol. IX, 1925, p. 267—274. A coloured reproduction of the portrait in: F. W. Stapel, *Geschiedenis van Ned. Indië,* Vol. V, p. 12.

SZÉKELY

Lived in Medan.
S. 14 *A young European frantically searching his shabby clothes for money, surrounded by his creditors.*
Water-colours. Signed (in the lower right-hand corner) Székely. 39 × 29 cm.
Note: Among the crowd there is a Chinese washerman(?), a Singhalese baker, and a native cobbler presenting his bill.

TICHELMAN, G. L.

Born Jan. 31, 1893, Palembang. Died Dec. 1961, Overveen.
He went to the Indies as a planter. In 1916 he joined the civil service. After repatriating he worked at the Royal Tropical Institute in Amsterdam as an assistant in the section for tropical products of the documentation office.

S. 11 *Thunder spirit.*
Oil on three-ply wood. 77,5 × 50 cm.
Note: Free composition representing the Thunder spirit according to the conception of that spirit by the Papuas of the Sepik region.

Provenance: Presented by Dr. P. Voorhoeve, 1962.

TIP, L.

Born Oct. 24, 1870, Culemborg. Died Febr. 10, 1953, Voorburg.
Civil service officer. Assistant resident at Pare-Pare, Bonthain, Bengkalis, Billiton and finally resident of Ternate. Retired in 1922.

K. 22, 1 *Ingang Kapoeas rivier* (Mouth of the Kapuas river).
Water-colours. Signed L. Tip. 18 × 23,5 cm.

K. 22, 2 *Overgang Tjamba rivier* (Crossing in the Tjamba river).
Ink. Signed LT. Dated 21-12-'43. 24,5 × 31,5 cm.
Note: The Tjamba river in S.W. Celebes.

K. 22, 3 *Zuid Sumatra* (South Sumatra).
Ink. 20,5 × 25,5 cm.
Note: Drawn between 1904—1911. The drawing shows pile-dwellings along the river.

K. 22, 4 *Bob* (The dog Bob).
Ink. Signed L. Tip. 18 × 13,5 cm.

K. 22, 5 *Reede Tobelo* (The roads of Tobelo).
Pencil on brown paper. Dated 1921. 19,5 × 28 cm.

K. 22, 6 *Teroes[an?]* (Terus[an?]).
Pencil. Dated 13/10-'16. 21 × 29,5 cm.
Note: The drawing shows houses with three landing-stages on the Bengkalis (?) river.

Provenance: Presented by Mrs. L. Tip-Suringa, widow of L. Tip, by courtesy of J. H. Maronier.

ULLMANN, L.

In Dec. 1849 he was second lieutenant of the Royal Dutch Indies Army; he was made lieutenant on Dec. 5, 1852, and captain in July 1857. On Dec. 15, 1852 he received a knighthood, the 'Militaire Willemsorde' 4th class. The Royal Intitute is in possession of two of Ullmann's manuscripts: H. 117, "Een blik in de Residentie Palembang 1855" with water-colours by Stroykens (see Stroykens); H. 400, "Het eiland Banka en zijne uitwendige gesteldheid" with water-colours by J. J. Geelhoed (see Geelhoed). These manuscripts are both dedicated to the commander of the East Indies Army, lieutenant-general jhr. F. V. A. Ridder de Stuers.

O. 11, 1 *Image near Palembang.*
Pencil. 14 × 22,5 cm.
Note: The first (A) illustration to the article "Hindoebeelden in de Binnenlanden van Palembang" (Hindu statues in the hinterland of Palembang), *Indisch Archief* 1850, Vol. 2, p. 493, is a reproduction of Ullmann's drawing.

References: A. N. J. Thomassen à Thuessink van der Hoop, *Megalithic Remains in South-Sumatra,* Zutphen 1932, p. 16—17: Tinggihari No. 6.

VER HUELL, Q. M. R.

Born September 11, 1787, Zutphen. Died 1860, Arnhem.
On his journeys as a naval officer he made use of his artistic talent to

record his impressions. He left a large body of work, now in the Municipal Museum in Arnhem and in the Maritime Museum in Rotterdam.

S. 17 *Poort van het Heilige Graf van Girie nabij Grissée op Java* (Gate of the holy tomb of Giri near Gresik in Java).
Water-colours. Signed (in margin) Q. M. R. Verhuell. 13,1 × 8,5 cm.
Note: Drawing with above description as well as the words "Het Volmaakte Geluk zij Uw Deel" (May perfect happiness be your destiny) and "Opregtheid" (Sincerity). Framed in passe-partout, 27 × 21,5 cm.

Provenance: Bought at auction, Van Stockum's antiquary, 1965.

References: H. J. de Graaf, "De tekeningen van de zee-officier Q. M. R. Ver Huell", *B.K.I.*, CXIII, 1957, pp. 261—273.

VERZIJL, G.

F. 12, 1—34 *Women's blouses and a tattoo-mark from Bada'.*
Water-colours. 34,5 × 49 cm. In portfolio, 57,5 × 40 cm.
Note: Thirty four paintings based on sketches and models of the late J. Woensdregt, one time missionary in Bada' in Central Celebes.

Provenance: Presented by Mrs. E. Woensdregt-Hoorweg, June 17, 1933.

References: N. Adriani en A. C. Kruyt, *De Bare'e sprekende Toradja's van Midden Celebes* (The Bare'e speaking Toradja's from Central Celebes), Vol. II, p. 218—219
J. Woensdregt, "Verloving en huwelijk bij de To Bada' " (Betrothal and marriage in the country of the To Bada's), *B.K.I.* LXXXV, 1929, pp. 245—290.

VORDERMAN, A. G.

Born Dec. 12, 1844, The Hague. Died July 12, 1902.
Physician in the Indies since 1866. Did important research especially on beri-beri and had an honorary doctorate conferred on him by Utrecht University.

K. 23 *Acht afbeeldingen van landverkenningen uit Straat Sunda, eenige dagen na de laatste eruptie van Krakatau; vervaardigd door A. G. Vorderman, Geneesheer, Bat. Havenwerken* (Eight pictures of landfalls in Sunda Strait, a few days after the last eruption of Krakatau: drawn by A. G. Vorderman, physician at the Bat. harbour-works).
Water-colours. 11 × 18 cm. In portfolio, 26,5 × 34 cm, with a printed title pasted on it. The drawings are mounted on pasteboard.

K. 23, 1 *Toppershoedje in het West-noordwest op 1 mijl* (Toppershoedje in the west-north-west, 1 mile's distance).

K. 23, 2 *Poeloe Merak ten Zuiden op 1¼ mijl* (Pulu Merak to the south 1¼ mile's distance).

K. 23, 3 *Dwars in den weg in het Zuid-zuidwest op 1.3/4 mijl* (The island of Dwars-in-den-Weg in the south-south-west 1.3/4 mile's distance).

K. 23, 4 *Krakatau. Piek in het Zuidwest half Zuid op 6 mijlen* (Krakatau. Peak in the south-west, 6 miles' distance).

K. 23, 5 *Sebessie in het West half Zuid op 3 mijlen* (Sebessie in the west, 3 miles' distance).

K. 23, 6 *Seboekoe op 1¼ mijl in het Westen* (Sebuko 1¼ mile to the west).

K. 23, 7a *Varkenshoek vóór den vloedgolf* (Varkenshoek before the tidal wave).

K. 23, 7b *Varkenshoek na den vloedgolf (Sumatra)* (Varkenshoek after the tidal wave (Sumatra)).
Note: Two paintings on one sheet.

K. 23, 8 *Onder de asch bedolven kampong Pangkoel aan den voet van den Radja-Bassa (Sumatra)* (Village of Pangkul buried by ahes at the foot of Mt. Rajah Bassa (Sumatra)).

References: R. D. M. Verbeek, *Krakatau, Album, bevatten 25 reproducties van aquarellen in kleuren der verwoeste gedeelten van Straat Soenda, genomen 2 maanden na de uitbarsting* .(These water-colours are different from those by Vorderman.)
Manuscript H. 565, Collection of papers of Dr. R. D. M. Verbeek relating to the eruptions of Krakatau.

WARDENAAR, Johan Willem Bartholomeus

Born 1785, Semarang. Died after 1864, probably Surabaya.
In 1798 he entered the naval college in Semarang, in 1801 he was a cadet engineer. Between 1805 and 1807 he was an ensign and assisted the lieutenant engineer H. C. Cornelius in surveying the Prambanan temples. In 1815 he drew a plan to scale of the antiquities of Majapahit. Apart from this he worked as a surveyor. He made a design for the mosque Kemayoran at Surabaya (transferred in 1848). In 1847 Dr. W. R. van Hoëvell visited him, and much admired Wardenaar's collection of antiquities, part of which came from "Djelok Toendo".

F. 13, 1 *Jalatuṇḍa, tomb and holy bathing place on the slope of mount Penanggungan.*
Water-colours. 39,5 × 25 cm.
Note: Drawn about 1815. Picture slightly damaged by insects. Pencil notes on reverse side. Reproduced in: F. D. K. Bosch and B. de Haan, "The Old Javanese Bathing Place Jalatuṇḍa", *B.K.I.* CXXI, 1965, pp. 189—232, pl. 2.

F. 13, 2—4 Drawings evidently intended for lithographing for the book *Reis over Java, Madura en Bali* by W. R. van Hoëvell, 3 vols., Amsterdam 1849—1854.

F. 13, 2 *Tjandi Brawoe. (Ruïnen van Modjopahit)* (Chandi Brahu. (Ruins of Majapahit)).
Pencil. Signed Wardenaar del. 22,5 × 15,5 cm.
Note: Lithographed in Van Hoëvell, *op cit.,* Vol. I, p. 176.

F. 13, 3 *Jalatunda, tomb and holy bathing place.*
Pencil. 22,5 × 15,5 cm.
Note: Lithographed in Van Hoëvell, *op. cit.,* Vol. II, p. 106. Copied from F. 13, 1, slightly different.

F. 13, 4 *Tjandi Djeddong* (Chandi Jedong).
Pencil. 22,5 × 15,5 cm.
Note: Lithographed in Van Hoëvell, *op. cit.*, Vol. III, p. 64.

WERDMÜLLER VON ELGG, P. A. C. H. T. H.

Was an infantry lieutenant during the campaign of Banjarmasin. He distinguished himself and received special mention in general orders (1863, no. 6). He made many drawings of landscapes and settlements in South Borneo. Some of these have been reproduced and published in two portfolio's: *Schetsen uit Bandjermasin* . . . uitgegeven tot een liefdadig doel bij H. M. van Dorp, 1862 (Sketches of Banjarmasin . . . published for a charitable cause by H. M. van Dorp, 1862) (20 lithographs); *Twaalf schetsen van belangrijke punten uit het rijk Bandjermasin in Borneo's Z. & O. kust naar de natuur geteekend* (Twelve sketches of important points in the sultanate of Banjarmasin, Borneo's south-east coast, drawn from nature). 1863.

O. 12, 1—6 *Talismans uit Bandjermasin (Talismans from Banjarmasin).*
Water-colours. 19,5 × 12,5 cm.

Note: Talismans originally owned by the pangerans (princes) Anta-Sari, Aminoela and Tonko-Brâhim, leaders in the Banjarmasin war, 1859—1863.

Provenance: Presented by I. Werdmüller von Elgg, then mayor of Culemborg. See attached letter dated Jan. 5, 1875 on reverse side of O. 12, 1.

WEIJERS, Berend Wolter

Born May 11, 1866, Kampen. Died May 9, 1949, The Hague.
Painter in the Hague, he trained at the academy of arts in that city.

S. 16 *Portrait of Jan Pieterszoon Coen* (knee-length).
Oil on canvas. 140 × 95 cm.
Note: This is the second copy based on the original in the museum in Hoorn. The first copy in possession of the Royal Institute was loaned

to the *Exposition Internationale Coloniale* in Paris in 1931. It was destroyed by the fire of the Dutch Pavilion. On the advice of Prof. Dr. W. M. Martin, director of the Mauritshuis, the commission for a new copy was given to B. W. Weyers.

UNKNOWN ARTISTS

S. 13 *D[irk] van Hogendorp op Novo Sion* (D[irk] van Hogendorp at Novo Sion).
Oil on canvas. To be dated between 1817 and 1822. 39 × 46 cm.

Description: Dirk van Hogendorp, living as an exile in Brazil since Febr. 10, 1817, is seated in front of his house on the small estate of Novo Sion, on the slope of the Corcovado near Rio de Janeiro, rented from the Englishman Will. Young. To the right in the foreground is a negro.

Provenance: Presented by Jkvr. M. A. van Hogendorp, see S. 12 (Suhr. C.).

★

K. 30 Ink and pencil drawings, probably intended as illustrations for a children's book about the life in the East Indies of a European family. The drawings are numbered 1—67 (several numbers occurring twice) and scales in cms. are indicated on all of them.

K. 30, 1 *Pointer.*
Pencil. 23 × 15 cm.

K. 30, 2 *Knife.*
Ink. 23 × 15 cm.

K. 30, 3 *Awl.*
Pencil. 15 × 23 cm.

K. 30, 4 *Tail-coat.*
Ink. 23 × 15 cm.

K. 30, 5 *Chin.*
Ink and pencil. 23 × 15 cm.

K. 30, 6 *A round case.*
Ink. 23 × 15 cm.

K. 30, 7 *Forefinger.*
Pencil. 15 × 23 cm.

K. 30, 8 *Dog.*
Pencil. 23 × 15 cm.

K. 30, 9 *Plug.*
Pencil. 15 × 23 cm.

K. 30, 10 *Drill.*
Ink. 23 × 15 cm.

K. 30, 11 *Spout of a tea-pot.*
Ink. 23 × 15 cm.

K. 30, 12 *Ball.*
Ink. 23 × 15 cm.

K. 30, 13 *Horse's foot.*
Pencil. 23 × 15 cm.

K. 30, 14 *Bonnet.*
Pencil. 23 × 15 cm.

K. 30, 15 *European boy holding a pup and running away from the pup's angry mother.*
Pencil. 23 × 15 cm.

K. 30, 16 *Plate with rice.*
Pencil. 23 × 15 cm.

K. 30, 17 *Bamboo bridge.*
Pencil. 16 × 25 cm.

K. 30, 18 *Man's back.*
Pencil. 23 × 17 cm.

K. 30, 19 *Pair of trousers.*
Pencil. 23 × 15 cm.

K. 30, 20 *European schoolboy.*
Pencil. 23 × 15 cm.

K. 30, 21 *Slipper.*
Pencil. 25 × 16 cm.

K. 30, 22 *Spear.*
Pencil. 25 × 16 cm.

K. 30, 23 *Bottle and liqueur glass.*
Pencil. 25 × 16 cm.

K. 30, 24 *Large capital I.*
Ink. 16 × 23 cm.

K. 30, 25 *European boy and girl sitting in the grass.*
Pencil. 23 × 17 cm.

K. 30, 26 *Hook.*
Ink. 15 × 23 cm.

K. 30, 27 *European boy with a rake in the garden.*
Pencil. 23 × 15 cm.

K. 30, 28 *European boy trying to catch his hat blown off by the wind wind.*
Pencil. 23 × 15 cm.

K. 30, 29 *Hook.*
Ink. 23 × 15 cm.

K. 30, 30 *Left foot.*
Pencil. 23 × 15 cm.

K. 30, 31 *Hen.*
Pencil. 23 × 15 cm.

K. 30, 32 *Feather.*
Ink. 23 × 15 cm.

K. 30, 33 *European boy in a rowing-boat.*
Pencil. 23 × 17 cm.

K. 30, 34 *Fish's head.*
Ink. 23 × 15 cm.

K. 30, 35 *Mouse and mousetrap.*
Ink. 25 × 16 cm.

K. 30, 36 *European girl sitting on a barrel with a cat and a doll.*
Pencil. 23 × 15 cm.

K. 30, 37 *Knife.*
Ink. 23 × 15 cm.

K. 30, 38 *The moon.*
Ink. 23 × 15 cm.

K. 30, 39 *Razor (old fashioned).*
Ink. 23 × 15 cm.

K. 30, 40 *Opened book.*
Ink. 25 × 16 cm.

K. 30, 41 *Native house.*
Ink. 23 × 17 cm.

K. 30, 42 *Native house.*
Ink. 23 × 16 cm.

K. 30, 43 *Bottle.*
Ink. 23 × 15 cm.

K. 30, 44 *Medecine bottle.*
Ink. 23 × 15 cm.

K. 30, 45 *Two European boys looking at a rose.*
Pencil. 23 × 15 cm.

K. 30, 46 *European woman playing with her child.*
Pencil. 23 × 16 cm.

K. 30, 47 *Two European girls looking at a cobweb.*
Pencil. 23 × 15 cm.

K. 30, 48 *European man in the garden giving a rose to a little girl.*
Pencil. 23 × 15 cm.

K. 30, 49 *Saw.*
Ink. 23 × 15 cm.

K. 30, 50 *European mother and child.*
Pencil. 23 × 16 cm.

K. 30, 51 *Thimble.*
Ink. 23 × 15 cm.

K. 30, 52 *A mouse eating ham.*
Pencil. 23 × 15 cm.

K. 30, 53 *European boy sitting on a rocking-horse.*
Pencil. 17 × 12 cm.

K. 30, 54 *Man in a buffalo cart.*
Pencil. 15 × 23 cm.

K. 30, 55 *Right eye.*
Ink. 23 × 15 cm.

K. 30, 56 *Spoon.*
Ink. 23 × 15 cm.

K. 30, 57 *Cat.*
Pencil. 15 × 23 cm.

K. 30, 58 *Sword.*
Ink. 15 × 23 cm.

K. 30, 59 *Nose.*
Pencil. 23 × 15 cm.

K. 30, 60 *Hammer.*
Ink. 15 × 23 cm.

K. 30, 61 *Stocking.*
Ink. 23 × 15 cm.

K. 30, 62 *Cable.*
Ink. 23 × 15 cm.

K. 30, 63 *Folded sunshade.*
Pencil. 23 × 15 cm.
Note: Pencil sketch of unfolded sunshade on reverse side.

K. 30, 64 *Cigar.*
Ink. 23 × 15 cm.

K. 30, 65 *Wood-pigeon.*
Pencil. 23 × 15 cm.

K. 30, 66 *European boy, sleeping with his toys.*
Pencil. 15 × 23 cm.

K. 30, 67 *Wall.*
Ink. 23 × 15 cm.

K. 30, 68 *Garden-boy at a native kitchen-range.*
Pencil. 23 × 15 cm.

K. 30, 69 *Gentleman's coat.*
Ink. 23 × 15 cm.

K. 30, 70 *European boy and gentleman.*
Pencil. 23 × 15 cm.

K. 30, 71 *Teacher with schoolboy.*
Pencil. 23 × 15 cm.

K. 30, 72 *Well.*
Ink. 23 × 15 cm.

K. 30, 73 *European girl chasing a butterfly.*
Pencil. 23 × 15 cm.

K. 30, 74 *European girl finding a knife.*
Pencil. 23 × 15 cm.

K. 30, 75 *A pair of scissors.*
Ink. 23 × 15 cm.

K. 30, 76 *Chinese boy joyfully leaving school.*
Pencil. 23 × 15 cm.

K. 30, 77 *Sheep.*
Pencil. 15,5 × 20 cm.
Note: A pencil sketch of a cat on reverse side.

K. 30, 78 *Shoe.*
Ink. 23 × 15 cm.
Note: A pencil sketch of a stocking on reverse side.

K. 30, 79 *Cat's head and dog's head.*
Pencil. 23 × 15 cm.
Note: A pencil sketch of a well on reverse side.

K. 30, 80 *Fish.*
Ink. 23 × 15 cm.

K. 30, 81 *European lady at a window.*
Pencil. 23 × 15 cm.

Provenance: Bought at Van Stockum's antiquary, The Hague, 1963.

★

GF. 1, 4 *Small river with outrigger proas, along the bank pile-dwellings and palms.*
Water-colours (in brown and yellow tints). 26,5 × 37 cm.

K. 27, 5 *Gezigt van 't Eilandje Polo Bato op de kust van Timor* (View of the island of Pulo Batu near the coast of Timor).
Pencil. 13 × 26 cm.
Note: Dated by Loose ca. 1845.

Provenance: GF. 1, 4 and K. 27, 5 bought at Loose's antiquary, The Hague, 1963.

★

K. 27, 3 *Kotta-Radja.*

Water-colours. Dated Decembris 1875. 15 × 23 cm.
Note: Oval, mounted on paper.

K. 27, 4 *Nagrak.*
Water-colours. Signed KS (or SK). Dated 15 Febr. '50. 16,5 × 28 cm.
Note: As there is a note on the reverse side saying "Salak '49", this Nagrak may be near Sukabumi.

Provenance: K. 27,3—4 bought at Vogel's second-hand bookshop, The Hague, 1960.

★

F. 14, 18 *View of Macassar.*
Water-colours. [Dated approx. 1840]. 17,2 × 31,5 cm.

Provenance: Bought June 1964 at the weekly market, Hobbemaplein, The Hague.

★

GF. 1, 5—8 Probably inspired by landscapes on Sumatra, but they are less realistic than fantastic.

GF. 1, 5 *Mountain chasm.*
Water-colours. 42,5 × 57,5 cm.

GF. 1, 6 *Lake between high cliffs.*
Water-colours. 42,5 × 50 cm.
Note: Rounded corners. In the middle of the painting are 2 soldiers.

GF. 1, 7 *View of a river with high rocks and some houses and trees.*
Water-colours. 41 × 55,5 cm.
Note: On lower left edge an indistinct signature: H. STISZ.S. (?).

GF. 1, 8 *View of a valley.*
Water-colours. 42,5 × 57,5 cm.

Provenance: Bought at Loose's antiquary, The Hague, 1963.

★

F. 14, 19—21 Three water-colours probably intended as illustrations in the books by H. J. Domis, *Reis over het eiland Java* (Journey across the island of Java), Pasoeroean 1829 and *De Residentie Pasoeroeang*, 's-Gravenhage 1836.

F. 14, 19 *Landscape with buffalo pulling a plow, Mt. Smeru in the background.*
Water-colours. 12 × 19,5 cm.

F. 14, 20 *Pasoeroeang.*
Ink and sepia. 9 × 26 cm.
Note: The drawing shows the former house of the resident in Pasuruan.

F. 14, 21 *Pasoeroeang.*
Ink and sepia. 9,5 × 21,5 cm.
Note: The drawing shows a road, houses with fenced yards, a mountain in the background.

Provenance: Bought 1963; from the estate of H. J. Domis, resident of Pasuruan in 1827—1831.

★

K. 27, 1 *Seascape with a large number of ships at Riau.*
Water-colours. 11 × 32,5 cm.
Note: In lower right corner "Riouw".

K. 27, 2 *Bay with ships, Riau.*
Water-colours. Dated (in lower right corner) Riouw 3 Junij [18]67. 11 × 32,5 cm.
Note: K. 27, 2 painted on reverse side of K. 27, 1.

Provenance: Bought at Loose's antiquary, The Hague, 1963.

★

K. 24 Series of 10 water-colours cut and mounted on Old Dutch, hand-made paper. All of them 26 × 17,8 cm. On the reverse side captions in writing in 18th century style.

K. 24, 1 *Boegineesen* (Bugis).

K. 24, 2 *Bugis.*

K. 24, 3 *Manges-tanga [en] Jap[ans] Kaki of Chineesche Vijg* (Mang-gistan and Japanese Kaki or persimmon).
Note: Indigenous fruits.

K. 24, 4 *Roode Ketasch-bloem. Huijs-Kikvors* (Red flower. Tree-frog).

K. 24, 5 *Vliegende Hagedis* (Draco volans).

K. 24, 6 *Tarnataans: Kobo; Batjangs: Dojong* (Sea-cow).

K. 24, 7 *Hollands: Zantkruijper; Japans: Kutschi; Maleijts: Gabbo Lauot* (Fish: Ophiocephalens striatus).

K. 24, 8 *Koning der Vliegende Visschen* (King of the flying fihs).
Note: Ptercis volitans? Indian Ocean.

K. 24, 9 *Mema* (Kind of fish).

K. 24, 10 *Calabemba* (Fish: Chelmon rostratus).

★

H. 468 Portfolio with contributions to the "Lapidarium der Oost- en West Indische bezittingen" (see Jaarverslagen Kon. Instituut 1888 and

1889). The sections on Banda and Amboina contain a number of drawings of Dutch antiquities. The latter have all been described in vol. I of *De Nederlandsche Oudheden in de Molukken* (Dutch antiquities in the Moluccas) by V. I. van de Wall, 's-Gravenhage 1928.

Ambon

a1 *Amboina, afbeeldingen van de Waterpoort Victoria, van de kerk en van verschillende grafzerken* (Amboina, pictures of the Victoria Watergate, the church and various tombstones).
Ink. 29 × 64,5 cm.
Note: See Van de Wall, pp. 137—142.

b1 *Fort "Wijk bij Duurstede" ter hoofdplaats Saparoea van de landzijde gezien* (Fort "Wijk bij Duurstede" in the principal town of Saparua as seen from the land side).
Chalk. 29 × 44,5 cm.
Note: See Van de Wall, pp. 196—199.

c1 *Fort Zeelandia te Haroekoe* (Fort Zeelandia in Haruku).
Pencil. 22,5 × 61 cm.
Note: See Van de Wall, pp. 202—203.

d1 *Fort Hoorn van de land- en van de zeezijde* (Fort Hoorn as seen from the land side and from the sea).
Chalk. 42 × 34,5 cm.
Note: See Van de Wall, p. 204.

e1 *Ruïne voormalig blokhuis Beverwijk te Sila (eil. Noesa Laut)* (Ruins of the former blockhouse Beverwijk in Sila (Isle of Nusa Laut)).
Chalk. 46 × 60 cm.
Note: See Van de Wall, pp. 206-207.

f1 *Grafzerken, wapenschilden in de protestantsche kerk te Amboina* (Tombstones, escutchens in the protestant church in Amboina).
Ink. Dated Amboina 1 Sept. 1890. 63 × 49 cm.

g1 *Gambar Kotta Middelburg Passo Afd. Amboina Pohon Pohon Waringin, ruïne* (Picture of fort Middelburg in Passo, district of Amboina. Banyan trees and a ruin).
Ink. 45 × 29 cm.
Note: On the same sheet: overgrowth with the same ruin. Pencil. Signed W. J. S. Nanlohij. Dated Tial-Serani 1 Nov. 1890. See Van de Wall, pp. 176—178.

Banda

a2 *Het Fort "Concordia" Groot Banda van de Noordzijde gezien* (Fort "Concordia", Greater Banda, as seen from the north).
Water-colours. 11 × 22,5 cm.
Note: See Van de Wall, pp. 100—107.

b2 *Het Fort "Revenge" op het eiland Ay, van de Noordzijde gezien* (Fort "Revenge" on the Isle of Ay, as seen from the north).
Water-colours. 12,5 × 23 cm.
Note: See Van de Wall, pp. 111—115.

c2 *Het Fort "Hollandia" te Groot Banda van de Z.Z. gezien* (Fort "Hollandia" in Greater Banda as seen from the south).
Water-colours. 11 × 22,5 cm.
Note: See Van de Wall, pp. 99—100.

d2 *Het Fort "Belgica" ter Hoofdplaats Neira van de N.W. zijde gezien* (Fort "Belgica" in the principal town of Neira, as seen from the north-west).

Water-colours. 15,5 × 22 cm.
Note: See Van de Wall, pp. 23—32.

e2 *Het Fort "Nassau" ter Hoofdplaats Neira van de Z.Z. gezien* (Fort "Nassau" in the principal town of Neira, as seen from the south).
Water-colours. 14 × 22,5 cm.
Note: See Van de Wall, pp. 21--23.

f2 *Overblijfselen van het Fort te Celaneon van de Noordzijde gezien* (Remains of the fort in Celaneon, as seen from the north).
Water-colours. 17,5 × 18,5 cm.
Note: See Van de Wall, p. 93: "geen spoor meer over" (nothing is left).

g2 *Overblijfselen van het Fort "Lakoeï" van de Z.Z. gezien* (Remains of fort "Lakui", as seen from the south).
Water-colours. 17,5 × 18 cm.
Note: See Van de Wall, p. 103: "Lackoey".

h2 *Het Fort "Kijk in de Pot" gelegen op den Vuurberg van de N.O. zijde gezien* (Fort "Kijk in de Pot" on the Vuurberg, as seen from the north-east).
Water-colours. 17 × 24,5 cm.
Note: See Van de Wall, pp. 70—71.

★

H. 583 A few drawings of agricultural implements are te be found in the manuscript of W. L. de Sturler i.a. concerning the exploitation of mineral wealth.

a *Ploughs, a rake and a hoe.*
Ink. 30 × 42 cm.

b *Sawah-ploeg* (Sawah-plough).
Ink. 14 × 20 cm.

Provenance: From the estate of W. L. de Sturler.

★

K. 25 Three water-colours drawn by pupils of the I.N.S. (Dutch-Indonesian-School) Moh. Sjafé'i in Kajutanam in Menangkabau. There is a caption on each drawing saying 'I.N.S. Kajutanam'.

K. 25, 1 *Landscape with lake in background (Lake of Manindjau?) Menangkabau houses in the foreground.*
Water-colours. 20,5 × 33,5 cm.

K. 25, 2 *Lake of Singkarah (?) with reflected clouds, ricefield in the foreground, mountains in the background.*
Water-colours. 22 × 32,5 cm.

K. 25, 3 *Mountain (Merapi?) with Menangkabau houses in the foreground.*
Water-colours. 21 × 32,5 cm.

Provenance: Presented by Mrs. S. C. L. Vreede-de Stuers, 1962.

★

K. 26 Various types of people of Sumatra's West-coast during the Padri-war, approx. 1830.

K. 26, 1 *Korporaal van de Javasche Barissan* (Corporal of the Javanese auxiliary troops).
Water-colours. 28,5 × 21,5 cm.
Note: A native soldier, barefoot, a sash round his waist, and on his head a kind of turban with a uniform-cap that is too large on top of it. He has shouldered his rifle from the barrel of which there is a Dutch flag hanging with some pseudo-Arabic characters on it. This is probably a soldier of the legion of Sentot, comrade in arms of Pangéran Dipa Negara who came over to the Dutch in 1827 and since then fought against the Padri's on Sumatra's West-coast. These soldiers were strict Mohammedans, observing the rules of their religion fairly closely, and covering their shaven heads with a turban.

K. 26, 2 *Klingalees* (Kling).
Water-colours. 28,5 × 19,5 cm.

K. 26, 3 *Doeblan (voorvechter van de maleische hulptroepen) op Sumatra* (Dublan (champion of the Malay auxiliary troops) in Sumatra).
Water-colours. 28,5 × 19,5 cm.
Note: Dubalang, Minangkabau word for the Malay *hulubalang.*

Provenance: Bought at B. M. Israel's second-hand bookshop, Amsterdam.

WEST INDIES

BORRET, Arnoldus Hyacinthus Aloysius Hubertus Maria

Born Oct. 28, 1848,Maastricht. Died Febr. 6, 1888.
He was self-taught and lived in Suriname. From 1878 on he was successive by registrar, member of the court of Justice and member of the Colonial Council. Until 1882 he made a large number of water-colours and drawings of landscapes and indigenous types. On Jan. 23, 1882 he joined the Redemptionists. In 1883 he became a priest and in 1887 vicar general and superior of the mission.

F. 1, 1 *Design for an album on Suriname.*
Water-colours. Signed AB. 20,5 × 32,5 cm.

F. 1, 2 *Indigenous types.*
Water-colours. Signed AB. Dated 20/8 1881. 27 × 37 cm.

F. 1, 3 *Indigenous types and every-day activities.*
Water-colours. Signed AB. Dated 15/3 81. 27 × 37 cm.

F. 1, 4 *Indigenous types and every-day activities.*
Water-colours. Signed AB. Dated 1881. 27 × 37 cm.

F. 1, 5 *Creole.*
Water-colours. 32 × 22 cm.

F. 1, 6 *Fort Nieuw Amsterdam.*
Pencil. 23,5 × 35 cm.
Note: Dated '78 on passe-partout.

F. 1, 7 *Plantation near a river.*
Pencil. 16,5 × 31 cm.

F. 1, 8 *St. Joseph's Orphanage.*
Pencil. 28,5 × 37 cm.

F. 1, 9 *Landscape with houses (of lepers?).*
Pencil. Dated: Paramaribo '79. 25 × 34 cm.

F. 1, 10 *Pad van Wanica* (Path of Wanica).
Pencil. Dated '81. 23 × 37 cm.

F. 1, 11 *Fisherman's hut near Batavia orphanage.*
Chalk. 25 × 35 cm.

F. 1, 12 *Verlaten Saramacca* (Abandoned [plantation] Saramacca).
Pencil. Dated Sept. '81. 27 × 37 cm.

F. 1, 13 *The plantation "Ma Retraite".*
Pencil. 24 × 36,5 cm.

F. 1, 14 *Bosch van plantage Livorno* (Forest of the plantation Livorno).
Pencil. Dated '79. 25 × 37 cm.

F. 1, 15 *Plantage Dijkveld* (The plantation „Dijkveld").
Pencil. Signed AB. Dated 1880. 27 × 37 cm.

F. 1, 16 *Plantage Livorno* (The plantation "Livorno").
Pencil. Dated '79. 25,5 × 34 cm.

F. 1, 17 *Pl. Catharina Sophia* (The plantation "Catharina Sophia").
Pencil. Dated 14 Octob. 78. 27 × 37 cm.

F. 1, 18 *Catharina Sophia.*
Pencil. Dated 22 Octob. 78. 35 × 25 cm.

F. 1, 19 *Sortie de bois. Plantage Catharina Sophia* (The edge of the wood. Plantation "Catharina Sophia").
Pencil. Signed AB. Dated '78. 34 × 27 cm.

F. 1, 20 *Weg Jagtlust* (Road [in the plantation] "Jagtlust").
Pencil. Dated 14/10 1881. 33,5 × 27 cm.

F. 1, 21 *Plantage Jagtlust* (The plantation "Jagtlust").
Pencil. Dated 22 Nov. 1881. 33,5 × 27 cm.

F. 1, 22 *Country road. Suriname.*
Pencil. 33 × 24,5 cm.

F. 1, 23 *De stoomboot Curaçao in quarantaine 5—26 Juli 1881* (The steamer Curaçao lying in quarantine 5th—26th July, 1881).
Pencil. 25 × 32 cm.

F. 1, 24 *Boats on a river.*
Pencil. 20 × 30 cm.

F. 1, 25 *Landing-place of Indian camp.*
Pencil. Signed AB. 25,5 × 35 cm.

K. 1, 1 *Band. Suriname.*
Pen and water-colours. 24 × 19,5 cm.
Note: Freely copied from plate No. 1 of *Costumes et types d'habitants de Suriname* dessinées par Th. Bray (See Bray K. 2, 1).

K. 1, 2 *Indian girl.*
Pencil. 18 × 14 cm.

K. 1, 3 *Kotomisi.*
Pencil. 18,5 × 14,5 cm.

K. 1, 4 *Jules Grami.*
Pencil. Dated Fort de France 21/5, 78. 23 × 14 cm.

K. 1, 5 *View from a country house.*
Ink and pencil. 19 × 25 cm.

Note: Presumably view of Government House and 'Tamarindelaan', Paramaribo.

K. 1, 6 *Creek.*
Pencil. 19 × 25 cm.

K. 1, 7 *River.*
Pencil. 18 × 25 cm.

K. 1, 8 *Indian camp.*
Pencil. 14 × 23 cm.

K. 1, 9 *Lazaret.*
Pencil. 13,5 × 20,5 cm.

K. 1, 10 *Cocosboom* (Coconut-tree).
Pencil. Dated Catharina Sophia 12 Octob. 78. 23 × 14 cm.

K. 1, 11 *Leproserie Batavia a/d Coppename* (Lazaret Batavia on the Coppename).
Pencil. Dated 3/9 1878. 14 × 23 cm.

K. 1, 12 *Saramacca (Plge. Catharina Sophia)* (Saramacca (Plantation Catharina Sophia)).
Pencil. Dated 9 Sept. 78. 14 × 23 cm.

K. 1, 13 *Forced rest during a river journey.*
Pencil. 14 × 23 cm.

K. 1, 14 *Bay.*
Pencil. 13 × 21 cm.

K. 1, 15 *St. Lucia.*
Pencil. Dated 23/5 78. 14 × 23 cm.

K. 1, 16 *Côte de la Martinique* (Coast of Martinique).
Pencil. Dated 22/5 78. 14 × 23 cm.

K. 1, 17 *Côte de la Martinique* (Coast of Martinique).
Pencil. Dated 23 Mai 78. 14 × 23 cm.

K. 1, 18 *Côte de la Martinique. Dominique au fond* (Coast of Martinique. Dominique in the background).
Pencil. Dated 22/5 78. 14 × 23 cm.

K. 1, 19 *Fort de France (Martinique).*
Pencil. Dated 21/5 78. 14 × 23 cm.

K. 1, 20 *Canal des Saintes* (Saints' Canal).
Pencil. Dated 20/5 78. 7½ du soir. 14 × 23 cm.

K. 1, 21 *(Guadeloupe) Canal des Saintes* ((Guadeloupe) Saints' Canal).
Pencil. Dated 7½h. du soir. 20 Mai 78. 14 × 23 cm.

K. 1, 22 *Pointe à Pitre (Guadeloupe).*
Pencil. Dated 20 Mai 78. 14 × 23 cm.

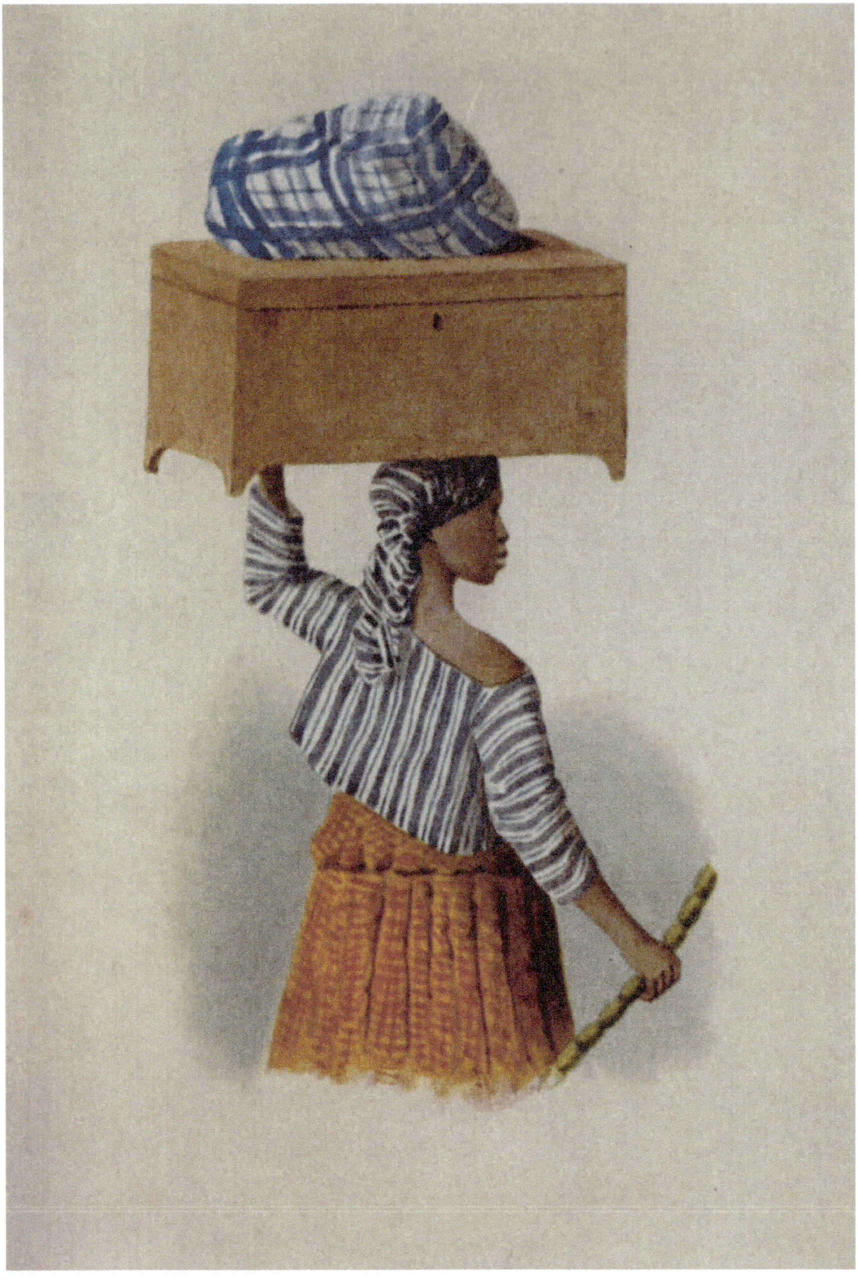

A. Borret A Surinam woman O. 1, 5

K. 1, 23 *(Guadeloupe) Pointe à Pitre.*
Pencil. Dated 20/5 78. 14 × 23 cm.

K. 1, 24 *St. Nazaire.*
Pencil. Dated 7 Mai 78. 12,5 × 21,5 cm.

K. 1, 25 *Pointe à Pitre (Guadeloupe).*
Pencil. Dated 20/5 78. 8h. du matin. 14 × 23 cm.

K. 1, 26 *Palm-tree.*
Pencil. Dated Catharina Sophia 15 Oct. 78. 23 × 14 cm.

K. 1, 27 *Quarantine station.*
Pencil. 14,5 × 24,5 cm.

K. 1, 28 *Plantage Catharina Sophia* (The plantation Catharina Sophia).
Pencil. Dated '78. 13 × 22,5 cm.

K. 1, 29 *Pointe à Pitre. Vue prise du balcon du Cercle — (Midi) —* (Pointe à Pitre, view from the balcony of the Cercle Midi).
Pencil. 20/5 78. 23 × 14 cm.

K. 1, 30 *Entrée de la baie de Trinidad* (Entrance to the bay of Trinidad).
Pencil. Dated 23/5 78. 14 × 23 cm.

K. 1, 31 *Forest path in Suriname.*
Pencil. 18 × 23,5 cm.

O. 1, 1 *Kotomisi.*
Water-colours. 15 × 9 cm.
Note: Published in *B.K.I.* CXI (1955), facing p. 4.

O. 1, 2 *Woman carrying a jar on her head.*
Water-colours. 15 × 9 cm.

O. 1, 3 *A watchman.*
Water-colours. Dated Catharina Sophia 21 Octob. 78. 15,5 × 11 cm.

O. 1, 4 *Native woman.*
Water-colours. 15 × 9 cm.

O. 1, 5 *Woman carrying a chest on her head.*
Water-colours. 17,5 × 10,5 cm.
Note: Published in *B.K.I.* CXI (1955), facing p. 1.

O. 1, 6 *Weeping woman.*
Water-colours. 18 × 13 cm.

O. 1, 7 *Fisherman.*
Water-colours. 15,5 × 9,5 cm.

O. 1, 8 *Woman with a tray on her head and a wooden bowl in her hand.*
Water-colours. 18,5 × 12,5 cm.

Note: Freely copied from Th. Bray (See Bray, K. 2, 2).

O. 1, 9 *Two women gossiping.*
Water-colours. 19 × 12,5 cm.

O. 1, 10 *A Hindustani.*
Water-colours. Dated pl. Catharina Sophia 17 Octob. 1878. 16 × 8 cm.

O. 1, 11 *Nursemaid, with European baby.*
Water-colours. 20 × 12,5 cm.

O. 1, 12 *Woman carrying child on her back and load on her head.*
Water-colours. 20 × 12,5 cm.
Note: Published in *B.K.I.* CXI (1955), between pp. 4 and 5.

O. 1, 13 *Indigenous woman.*
Water-colours. 17,5 × 10,5 cm.

O. 1, 14 *Oarsman.*
Water-colours. 18 × 12,5 cm.
Note: Freely copied from Th. Bray (See Bray, K. 2, 21).

O. 1, 15 *Flying a kite.*
Water-colours. 15,5 × 8 cm.

O. 1, 16 *Kotomisi.*
Water-colours. 16,5 × 11 cm.

Note: Published in *B.K.I.* CXI (1955), between pp. 4 and 5.

O. 1, 17 *Kotomisi.*
Water-colours. 21 × 13 cm.

O. 1, 18 *Indigenous woman.*
Water-colours. 19,5 × 12,5 cm.
Note: Freely copied from Th. Bray (See Bray, K. 2, 7).

O. 1, 19 *Man from Suriname.*
Ink. 15,5 × 9 cm.
Note: With caption "naar Th. Bray" (copied from Th. Bray) (See Bray, K. 2, 10).

O. 1, 20 *Leper woman.*
Pencil. Dated Batavia (leproserie) 4 Sept. 1878. 14,5 × 9,5 cm.

O. 1, 21 *Old man.*
Ink. 15 × 8,5 cm.
Note: Freely copied from Th. Bray (See Bray, K. 2, 3).

O. 1, 22 *Resting woman.*
Ink. 13,5 × 9,5 cm.

O. 1, 23 *Indian girl.*
Pencil. Dated "Indiaansch kamp a/d Coppename" (Indian camp on the Coppename) 4 Sept. 1878. 20 × 12 cm.

O. 1, 24 *Man's head.*
Pencil. Dated 16 Octob. 78. 18,5 × 12,5 cm.

O. 1, 25 *Three women.*
Water-colours. 12 × 19 cm.

O. 1, 26 *Three women.*
Water-colours. 12 × 18,5 cm.

O. 1, 27 *Woman and child.*
Ink. 11 × 17 cm.

O. 1, 28 *View of river in the mountains.*
Pencil. 12 × 16,5 cm.

O. 1, 29 *Plantage Catharina Sophia* (Plantation Catharina Sophia).
Pencil. Dated '78. 10,5 × 14 cm.

O. 2, 1 *Portrait of a woman.*
Water-colours. 12 × 9,5 cm.

O. 2, 2 *Man.*
Washed ink. 11,5 × 7 cm.
Note: With caption "naar Th. Bray" (copied from Th. Bray) (See Bray, K. 2, 3).

O. 2, 3 *Headcloth.*
Water-colours. 5,5 × 4 cm.

O. 2, 4 *Creole.*
Water-colours. 4,5 × 4 cm.

O. 2, 5 *Woman's head.*
Water-colours. 5,5 × 4,5 cm.

O. 2, 6 *Woman's head.*
Water-colours. 8 × 5,5 cm.

O. 2, 7 *Head-cloth of a Kotomisi.*
Water-colours. 5 × 4 cm.

O. 2, 8 *Woman.*
Water-colours. 11,5 × 6 cm.

O. 2, 9 *Woman carrying a load on her head and a child on her hip.*
Water-colours. 10 × 6 cm.

O. 2, 10 *Man with a stick.*
Water-colours. 11,5 × 6 cm.

O. 2, 11 *Kotomisi head-dress as seen from behind.*
Water-colours. 9,5 × 7 cm.

O. 2, 12 *Woman with head-cloth.*
Water-colours. 11 × 9 cm.

O. 2, 13 *Flying kites.*
Water-colours. 12,5 × 10 cm.

O. 2, 14 *Head-cloth.*
Water-colours. 10,5 × 10 cm.

O. 2, 15 *Man sitting.*
Water-colours. 9,5 × 7 cm.

O. 2, 16 *Portrait study.*
Water-colours. 11 × 8 cm.

O. 2, 17 *Pipe-smoker.*
Ink. 11,5 × 7 cm.
Note: With caption "naar Th. Bray" (copied from Th. Bray) (See Bray, K. 2, 4).

O. 2, 18 *Man's head.*
Washed ink. 6,5 × 4,5 cm.

O. 2, 19 *Woman's head.*
Washed ink. 5,5 × 3,5 cm.

O. 2, 20 *Man carrying a bottle on his head.*
Washed ink. 7 × 5 cm.

O. 2, 21 *Kotomisi.*
Ink. 7 × 5,5 cm.

O. 2, 22 *Kotomisi.*
Ink. 6,5 × 4 cm.

O. 2, 23 *Boy carrying jugs.*
Ink. 8,5 × 5 cm.

O. 2, 24 *Head of a Kotomisi.*
Ink. 6 × 4,5 cm.

O. 2, 25 *Kotomisi with a load on her head.*
Ink. 6,5 × 4,5 cm.

O. 2, 26 *Man from India.*
Ink. 9 × 6,5 cm.

O. 2, 27 *Chinese merchant.*
Ink. 5 × 4,5 cm.

O. 2, 28 *Fish-seller.*
Ink. 3 × 5 cm.

O. 2, 29 *Man with hat.*
Ink. 4 × 5 cm.

O. 2, 30 *Creole with fringe of beard.*
Ink. 5 ×3,5 cm.

O. 2, 31 *Woman sitting.*
Ink. 7,5 × 5 cm.

O. 2, 32 *Man's head.*
Ink. 5 × 4 cm.

O. 2, 33 *Man from India.*
Ink. 6 × 5,5 cm.

O. 2, 34 *Boy's head.*
Ink. 6,5 × 6 cm.

O. 2, 35 *Woman carrying a load on her land.*
Ink. 7 × 4,5 cm.

O. 2, 36 *Man with pointed hat.*
Ink. 7 × 5 cm.

O. 2, 37 *Man's head.*
Ink. 7 × 5 cm.

O. 2, 38 *Woman with hat and head-cloth.*
Ink. 6,5 × 5 cm.

O. 2, 39 *Woman with a stick.*
Ink. 6 × 4 cm.

O. 2, 40 *Sketch of a woman.*
Ink. 6 × 4,5 cm.

O. 2, 41 *Kotomisi dancing.*
Ink. 9 × 6 cm.

O. 2, 42 *Branch as head-load.*
Ink. 5,5 × 8 cm.

O. 2, 43 *Kneeling woman.*
Ink. 7,5 × 11 cm.

O. 2, 44 *Woman with branch.*
Ink. 10,5 × 5 cm.

O. 2, 45 *Kotomisi.*
Ink. 8 × 5,5 cm.

O. 2, 46 *Kotomisi crossing a river.*
Washed ink. 9,5 × 5 cm.

O. 2, 47 *Kotomisi.*
Ink. 10 × 6,5 cm.

O. 2, 48 *Woman sitting.*
Ink. 8 × 6 cm.

O. 2, 49 *Man leaning against a tree.*
Ink. 9 × 5 cm.

O. 2, 50 *Bathing woman.*
Ink. 8 × 4,5 cm.

O. 2, 51 *Man carrying load on his head.*
Ink. 9 × 5 cm.

O. 2, 52 *Two head-loads.*
Ink. 11 × 6,5 cm.

O. 2, 53 *Man with bundle on his head.*
Ink. 12 × 6 cm.
Note: With caption "naar Th. Bray" (copied from Th. Bray) (See Bray, K. 2, 6).

O. 2, 54 *Woman from India.*
Ink. 12 × 6 cm.

O. 2, 55 *Woman from India.*
Ink. 12 × 6,5 cm.

O. 2, 56 *Boschneger* (Bush Negro).
Pencil. 9,5 × 7,5 cm.

O. 2, 57 *Indian.*
Ink and pencil. Dated 16 Octob. 78. 10,5 × 9 cm.

O. 2, 58 *Man with head-cloth and hat.*
Pencil. 11,5 × 8,5 cm.

O. 2, 59 *Man with plaited hair.*
Ink. 13 × 9 cm.

O. 2, 60 *Large head-load.*
Ink. 7 × 11 cm.

O. 2, 61 *Horse-cart.*
Ink. 8,5 × 12,5 cm.

GF. 1, 1 *Design for a title-page.*
Pencil. Signed AB. 64 × 47 cm.
Note: Text of the title-page: "Suriname. Gezichten, typen en costumen; naar de natuur geteekend door A. Borret" (Suriname. Scenery, types and costumes; drawn from nature by A. Borret). With ornamental border with bird of prey and basket of fruit.

GF. 1, 2 *Design for a table of contents.*
Pencil. Signed AB. 64 × 47 cm.
Note: With border of liana's with monkeys, with text: "Aanbieding van het souvenir. Paramaribo 15 Januari 1882" (Presentation of the souvenir. Paramaribo January 15, 1882).

Provenance: Presented by Th. L. M. H. Borret, Sept. 1911.

Reference: J. H. Maronier, "Surinaamse aquarellen en tekeningen van A. Borret", *B.K.I.* CXI (1955), pp. 1—5

BRAY, Th.

Born approx. 1814.
Planter in Surinam, of French origin. Became manager of „Vossenberg" in 1844 and was manager-owner of "Spieringshoek" from 1859—1861. Went to Holland in 1862 but returned to Surinam in 1868. A number of his drawings is in possession of the Royal Tropical Institute in Amsterdam.

K. 2 *Sketches of Surinam and its people.*
25 lithographs, the original drawings of which are kept in the Koloniale Bibliotheek (Colonial Library) of Paramaribo. The lithographs were printed by Petit, Paramaribo, 1850. A French edition is also in existence. To the former publication belongs a list of explanations of the pictures ("explicatie van de plaaten") to be found in the collection "Atlas van Stolk" in Rotterdam. All lithographs are hand-coloured. 22,5 × 30,5 cm. They are numbered 1—25.

K. 2, 1 *Orkest van een baljaar partij (Neger dans)* (Band of a 'baljaar partij' (Negro dance)).
Signed Th. Bray (mirror writing).

K. 2, 2 *Koek rondventster* (Woman peddling cake).
Signed Th. Bray.

K. 2, 3 *Neger familie* (Negro family).

K. 2, 4 *Veildmeid en kinder oppasster (Kreoolen mama)* (A woman and a children's nurse (Creole mother)).
Signed Th. Bray.

K. 2, 5 *Caraiben Indianen* (Caribbean Indians).
Signed Th. Bray.

K. 2, 6 *Plantaadje negerinnen* (Plantation negresses).
Signed Th. Bray.

K. 2, 7 *Plantaadje Directeur met huismeid en huisjongen* (Manager of a plantation with housemaid and houseboy).
Signed Th. Bray.

K. 2, 8 *Negerin* (Negress).
Signed Th. Bray.

K. 2, 9 *Stadbewoners* (Town's people).
Signed Th. Bray.

K. 2, 10 *Groep van plantaadje negers* (Group of plantation negroes).
Signed Th. Bray.

K. 2, 11 *Heiden begrafenis op plantaadje* (Heathen burial on a plantation).
Signed Th. Bray.

K. 2, 12 *Kappen en vervoeren van suikerriet* (Cutting and transportation of sugar-cane).
Signed Th. Bray.

K. 2, 13 *Directeur met vroedvrouw en Kraamvrouw* (Manager with midwife and new mother).
Signed Th. Bray.

K. 2, 14 *Plantaadje huismeid* (Plantation housemaid).
Signed B.

K. 2, 15 *Administrateurs' klerk weekgeld uitbetalende aan kraam & zwangere vrouwen* (Manager's clerk paying weekly wages to a new mother and a pregnant woman).
Signed Bray.

K. 2, 16 *Kostwachter voor zijne hut* (Watchman of vegetable-fields in front of his cabin).
Signed Th. Bray.

K. 2, 17 *Plantaadje negerhuizen* (Negro dwellings on a plantation).
Signed Th. Bray.

K. 2, 18 *Kleurling Misie* (Coloured woman).

K. 2, 19 *Stad slavinnen* (Woman slaves from town).
Signed Th. Bray.

K. 2, 20 *Bezoek van den directeur in het slaven hospitaal* (Manager visiting the hospital for slaves).
Signed Th. Bray.

K. 2, 21 *Boschnegers* (Bush negroes).
Signed Th. Bray.

K. 2, 22 *Plantaadje huismeid in dans Kleeding* (Plantation housemaid in dancing-costume).
Signed Th. Bray.

K. 2, 23 *Arowakke Indianen* (Arawak Indians).
Signed Th. Bray (mirror writing).

K. 2, 24 *Plantaadje negerin in dans Kleeding* (Plantation negress in dancing-costume).
Signed Th. Bray.

K. 2, 25 *Suikerriet Molen* (Sugar-mill).
Signed Th. Bray.

References: Th. Bray, "Uit het dagboek van een Surinaamsche planter" (Utr. Prov. en Stedelijk Dagblad; after 1863). The numbers K. 2, 3 and K. 2, 21 have been published in *Cultureel Indië,* Vol. V, pp. 258—259.

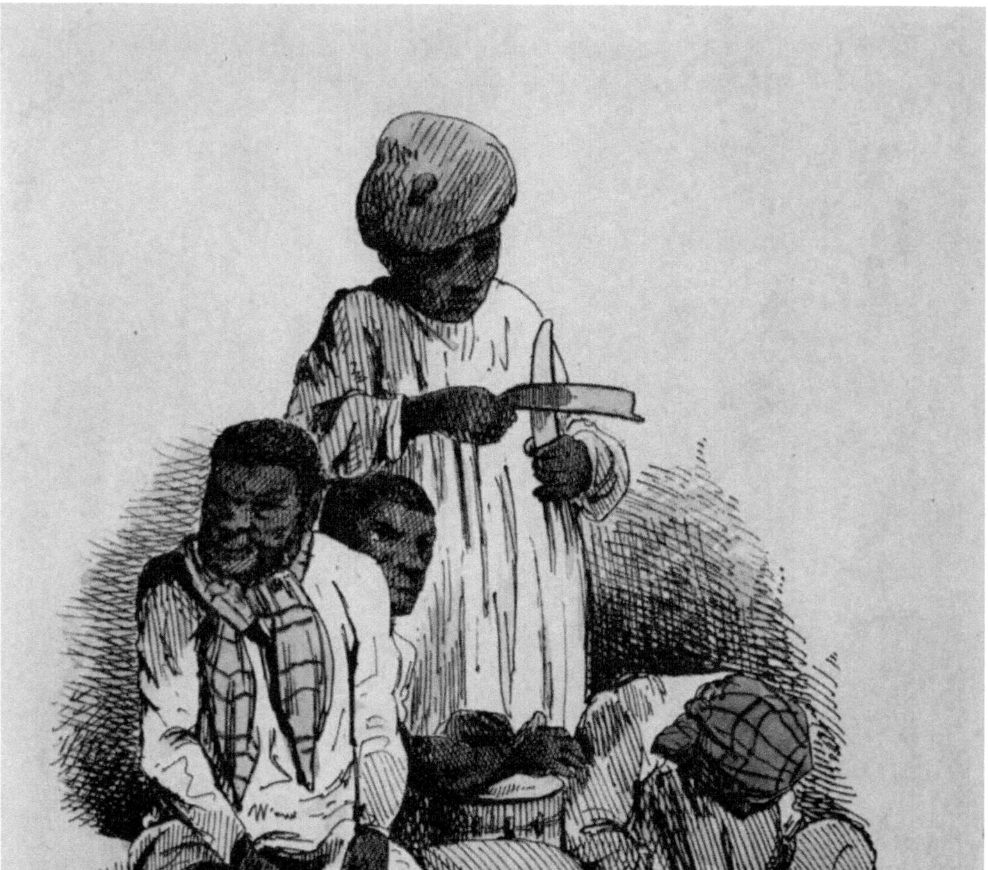

TH. BRAY Negro band, Surinam K. 2, 1

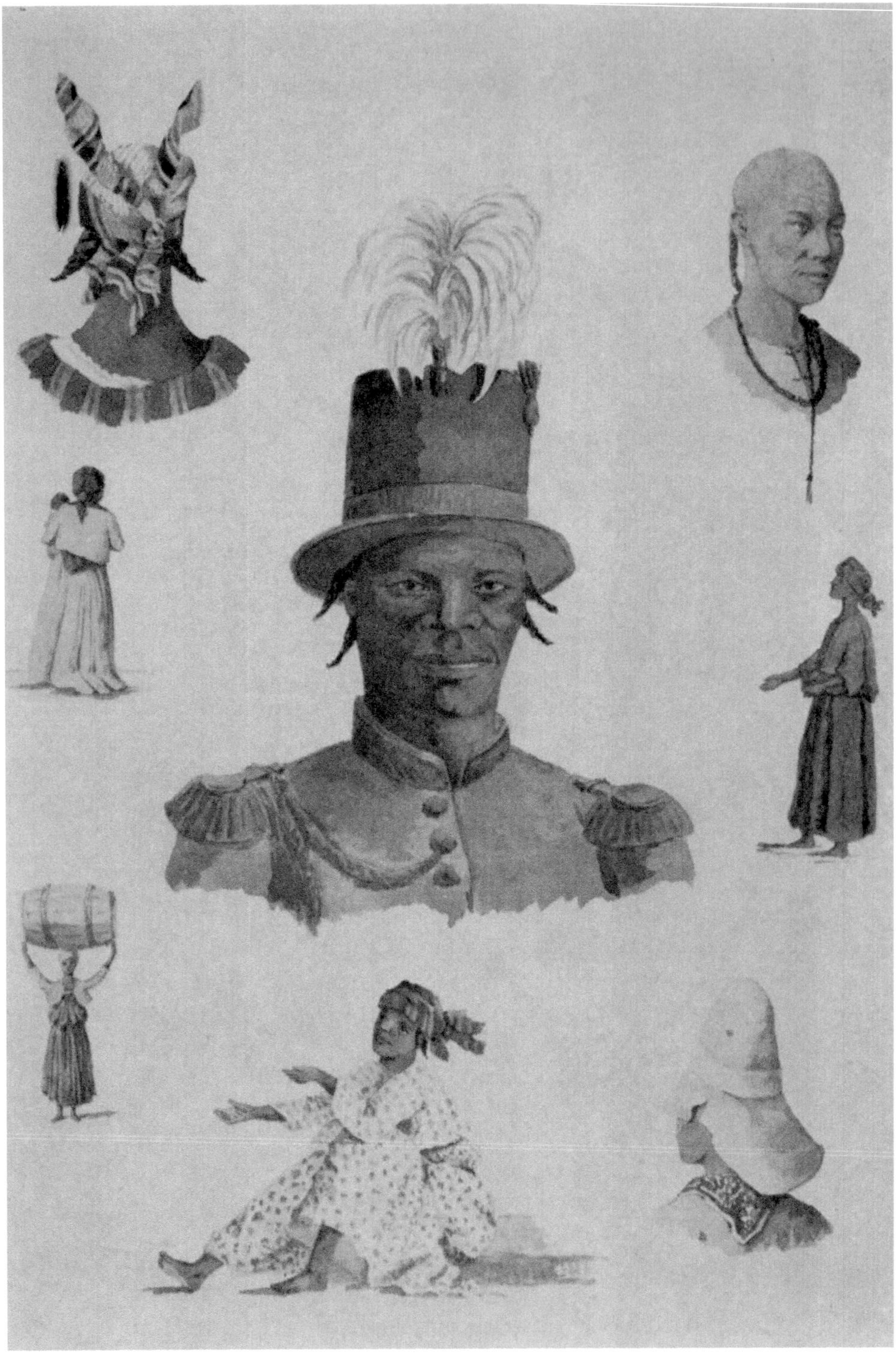

A. BORRET Indigenous types, Surinam F. 1, 3

MARKÉE, Cornelis

Painter and cartographer in Middelburg, Zeeland, about 1760. He made a compilation of all the orders of knighthood, containing 144 coloured pictures, in 1759. Other works by Markée are: a water-colour painting titled "Gezicht op St. Laurens" and dated 1767, now in the museum of the Zeeuwsch Genootschap in Middelburg (cf. W. S. Unger, *Catalogus van den historisch-topographischen atlas van het Zezuwsch Genootschap van Wetenschappen,* Middelburg 1931—1957, Vol. II, No. 956) and maps of Zeeland and Walcheren, now in the Provincial Archives. Cf. F. Nagtglas, *Levensberichten van Zeeuwen,* Vol. II, Middelburg 1893, p. 135.

F. 8 *Verzameling van voorscheijden uijtlansche gewasse als insecktens, alles seer naaukeurigh nagemackt en daar ik de rupsen niet van in wist, heb ik deselve op het een of ander gewas geplaast om an de wondere Gods te gedenken, hoe hij alles geschaapen heeft. Terneergesteldt binne Middelburg in voorscheijde jaaren.* 1763 (Collection of various exotic plants and insects, all very accurately copied and whenever I did not know the caterpillar I placed the butterfly on some plant or other, in order to be mindful of the wonders of God, the way He created everything. Put on paper in the town of Middelburg over a period of several years. 1763).
Mostly water-colours. All sheets 37,5 × 48 cm.
Note: The parchment bound album contains 83 sheets, covered on one side with coloured drawings of butterflies, cocoons and caterpillars with plants and fruits. Most of the drawings are freely copied from the work of Maria Sybilla Merian as published in her *Metamorphosis Insectorum Surinamensium* (1st ed. 1705), some from other, unidentified works and a few from nature. The rather detailed captions have been abbreviated here. The references to Merian's book state the numbers of the plates from which Markée's pictures have been copied.

F. 8, 1 *Title-page with border in rococo style.*
Chalk.

F. 8, 2 *Dese rups is uijt de aarde gespiet en dese uijl uijt voortgekomen* (This caterpillar has been dug up from the earth and has turned into this butterfly).

F. 8, 3 *Een Tack Nieswortel* (Branch of hellebore).

F. 8, 4 *Een Tack Kanella* (Branch of 'kanella').

F. 8, 5 *Orange.*
Cf. Merian LII.

F. 8, 6 *Branch of a lemon tree.*
Cf. Merian XVII.

F. 8, 7 *Distel* (Thistle).
Cf. Merian XXIV.

F. 8, 8—9 *Distel genaamt maccaj* (Thistle called 'maccaj').
Note: Cf. Merian VI. The Surinam word for thistle is *maká*.

F. 8, 10—11 *Een Tack van een Anena* (Branche of an 'anena').
Note: Markée says: Dese cappel was berustende bij de heer R. Winthuysen en daar van te leen gehadt om na te maken (this butterfly was part of the collection of R. Winthuysen and it was lent to me in order to make this sketch).

F. 8, 12 *Koffi boone tack* (Branch of a coffee-shrub).

F. 8, 13 *Indego tack* (Branch of indigo).

F. 8, 14 *Tackje van een veijgenboom* (Twig of a fig-tree).
Cf. Merian LXVII.

F. 8, 15 *De verandering van een uijtlansche grasrups* (The metamorphosis of an exotic caterpillar).
Cf. Merian LXIII.

F. 8, 16 *Cappel bij de heer Daniel Luycx Massis berustende* (Butterfly in the collection of Mr. Luyxc Massis).
Note: Mr. Luyxc Massis was one of the managers and principal shareholders of the Dutch West-Indian Co.

F. 8, 17 *Lantare draager* ((Lightgiving insect popularly called) 'lantern bearer').
Cf. Merian XLIX.

F. 8, 18 *Wijngaart rank* (Vine tendril).
Cf. Merian XXXIV.

F. 8, 19 *Branch of a fig-tree.*
Cf. Merian XXXIII.

F. 8, 20—21 *Een tack van een Limmetyens boom* (Branche of a lemon-tree).
Cf. Merian XVII.

F. 8, 22—23 *Water Meloenen* (Water-melons).
Cf. Merian XV.

F. 8, 24—25 *Caschou boome Tack* (Branch of a cashew-tree).
Cf. Merian XVI.

F. 8, 26—27 *Een Tack van de Guaiaves boom* (Branch of the 'guaiave'-tree).
Cf. Merian XIX.

F. 8, 28—29 *Een Tack van een Jasmijnboom* (Branch of a jasmin-shrub).
Cf. Merian VIII.

F. 8, 30—31 *Een Tack van een Veijgeboom* (Branch of a fig-tree).
Cf. Merian LXVII.

F. 8, 32—33 *Een Tack van een Cattoen boom* (Branch of a cotton-tree).
Cf. Merian X.

F. 8, 34 *Een tweede soort van Rups op de Cattoen boom* (A second kind of caterpillar on the cotton-tree).
Cf. Merian X.

F. 8, 35 *Dese Cappel heb ik wel onder Mijn Versameling dog de Rups is mijn daarvan onbekendt* (I have this butterfly in my collection, but its caterpillar is unknown to me).
Note: The pictures of both butterflies have clearly been copied from Merian XXXVI; the latter however also contains the drawing of a caterpillar.

F. 8, 36—37 *Gewas in Suernaamen okkerum, anders Althea genaamt* (plant called 'okkerum' or 'althea' in Surinam).
Cf. Merian XXXVII.

F. 8, 38—39 *Een Tack Van Een Mispelboom* (Branch of a medlar-tree).
Cf. Merian LIII.

F. 8, 40—41 *Een Tack Van de Gitte Gom* (Branch of gamboge).

F. 8, 42—43 *Pompelmoes* (Grape-fruit).
Cf. Merian XXIX.

F. 8, 44—47 *Mannier in leven der kakkerlacken op Aneas* (How the cockroaches live on pineapple).
Cf. Merian I—II.

F. 8, 48—49 *Gernaat* (Pomegranate).
Cf. Merian IX.

F. 8, 50—51 *Manihot of mayot. En wort van dese wortel het Cassavie brood gemackt* (Manioc. From this root cassave-bread is made).
Cf. Merian IV and V.

F. 8, 52—54 *De Pellesaden boom* (Palissade- or euterpe-palm).
Cf. Merian XI.

F. 8, 55 *Pruijmeboom* (Plum-tree).
Cf. Merian XIII.

F. 8, 56 *Banana.*
Cf. Merian XII.

F. 8, 57 *Wilde Cetroenen* (Wild lemons).
Cf. Merian XXVIII.

F. 8, 58—59 *Roosen* (Roses).
Cf. Merian XXXI.

F. 8, 60—61 *Kleijne Zuurzak* (Little sour-sop).
Cf. Merian III.

F. 8, 62—63 *Groote Zuurzak* (Large sour-sop).
Cf. Merian XIV.

F. 8, 64 *Een tweede soort van Rups op de Zuurzak* (A second kind of caterpillar on the sour-sop).
Cf. Merian XIV.

F. 8, 65 *Een Capelle dog de Rups is onbekent* (A butterfly but the caterpillar is unknown).

F. 8, 66—67 *Amirikaansche Kerschen* (American cherries).
Cf. Merian VII.

F. 8, 68 *Roode Lelie* (Red lily).
Cf. Merian XXII.

F. 8, 69 *Een Gras Rups met sijn verandering* (A caterpillar and its metamorphosis).
Cf. Merian XXII.

F. 8, 70—71 *Baccovis, met de verandering van de Rups tot de Cappel die de kleijne atlast Genaamt wordt* (Bacove (a kind of banana) and the metamorphosis of the caterpillar into the butterfly called the little atlas).
Cf. Merian XXIII.

F. 8, 72—73 *Een tack Banelle* (Branch of vanilla).
Cf. Merian XXV.

F. 8, 74 *Cacau boom* (Cocoa-tree).
Cf. Merian XXVI.

F. 8, 75 *Zodoms blad* (Leaf of Solanum).
Cf. Merian XXVII.

F. 8, 76—79 *Een tweede, derde, vierde soort van Rups met de verandering op de Guiaves boom* (A second, third and fourth kind of caterpillar and their metamorphosis on the 'guiave'-tree).
Cf. Merian XIX and LXI.

F. 8, 80—81 *Een tack Van den Gummi gutta boom* (Branch of the caoutchouc tree (Hevea Guianensis)).
Cf. Merian XX.

F. 8, 82—83 *Marquiaas boom of Passie boom* (Marquisa tree or passion-tree).
Cf. Merian XXI.

UNKNOWN ARTISTS

H. 451 In 1888 the Royal Instute made a request to the governor of Surinam to collect data concerning important buildings, churches, plantations, etc. (see Jaarverslagen 1888—1889). These data, among which are some water-colours and pencil drawings, were compiled in manuscript H. 451.

a *Wapenbord op de plantage "Rust en Werk" in het District Beneden Cottica* (Hatchment of the plantation "Rust en Werk" in the District of the Lower Cottica).
Water-colours. 16,5 × 20 cm.
Note: The drawing shows the front of the hatchment.

b *Wapenbord op de plantage "Rust en Werk" in het District Beneden Cottica* (Hatchment of the plantation "Rust en Werk" in the District of the Lower Cottica).
Water-colours. 16,5 × 20 cm.
Note: The drawing shows the reverse side of the hatchment; in ornamental border a motto "Mits men hier landwaerts rust en egter vordert 't werk. Dat het Gemeen vereyst soo noemt men 't Rust en Werk. AN 1760 NO".

The original hatchment is part of the collection of the Royal Tropical Institute, Amsterdam; see Catalogue of the exhibition "Wonen in de wijde wereld" of the Foundation for the cultural history of overseas Dutch, Dec. 20, 1963—March 16, 1964; Cat. No. 286.

c *Wit marmeren steen vóór de stoep van plantage "de Eendragt" in het district Perica. "De Dageraat". Deeze plantagie door den Edelgestrengen Heer G. A. D. de Graaff den 7den May 1767 begonnen en in den jaare 1786 tot een suikerplantagie aangelegd* (White marble stone in front of the steps of the plantation "de Eendragt" in the district of Perica. "De Dageraat". This plantation was set up by the hon. G. A. D. de Graaff on May 7, 1767, and was laid out as a sugar plantation in 1786).
Pencil. 20 × 33 cm.

d *Grafzerk op de plantage "Alkmaar" in het district Beneden Commewijne* (Tombstone on the plantation "Alkmaar" in the district of Lower Commewijne).
Pencil. 21 × 33,5 cm.
Note: With inscription "Charles Godefroy geb. 10 Mij 1704, ofer. 9 Juny 1773".

★

F. 14, 22 *View of Paramaribo from the river, with sailing-ship and state-barge.*
Water-colours. Approx. 1850. 14,5 × 32,7 cm.
Provenance: Bought at the weekly market, Hobbemaplein, The Hague, 1964.

CEYLON

KALENBERG VAN DORT, John Leonard

Born July 28, 1831. Died March 24, 1898.
Self-taught, designed illustrations and caricatures for English periodicals i.a. "The Graphic".

O. 6 *"Reminiscences of the Dutch occupation of the Island of Ceylon* from sketches of existing or extinct remains of Forts, Buildings, Canals, Inscriptions, Tombstones, etc., within or in the vicinity of some of the principal towns in Ceylon by John Leonard Kalenberg Van Dort of Colombo, Ceylon 1888—9. Prepared at the request of P. Daendeliker, Esq. Consul for the Netherlands at Colombo, Ceylon". (At the request and expense of the Royal Institute).

49 water-colours with title-page bearing the above description, in portfolio 25 × 20 cm. All of these have English captions and are signed J. L. K. Van Dort and dated 1888 or 1889.

O. 6, 1—3 *Three views of the old fort of Colombo in 1848.*

O. 6, 1 *From the sea.*
5 × 34 cm.

O. 6, 2 *From Galle Face.*
5 × 34 cm.

O. 6, 3 *From the Lake.*
5 × 34 cm.

O. 6, 4 *A copy from the oldest map of the Fortress of Colombo extant.*
18,5 × 13 cm.

O. 6, 5 *The Galbokka Gate (stone Quarry Gate) a sole remaining relic in entirety of the Dutch Fort of Colombo.*
7 × 12,5 cm.
Note: On the gate appears the date 1676.

O. 6, 6 *Date on wall of old Dutch Store Vaults, in the Fort of Colombo.*
8 × 5,5 cm.
Note: On the wall appears the date 1707.

O. 6, 7 *Water Gate in the Fort of Colombo (now closed up).*
8 × 5,5 cm.
Note: On the gate appears the date 1676.

O. 6, 8 *The old Belfry at Kayman's Gate. Colombo.*
18 × 13,5 cm.

O. 6, 9 *The Governor's official Residence in the Fort of Colombo. During the Dutch occupation of Ceylon.*
8 × 13 cm.

O. 6, 10 *Entrance porch to the Governor's residence in the Fort of Colombo.*
8 × 13 cm.

O. 6, 11 *The old Dutch House near the General Cemetry, in the Marandahn Cinnamon Gardens. Colombo. The residence of the Superintendant of the gardens under the Dutch Government.*
17 × 13 cm.

O. 6, 12 *Double monogram (G.F.V.B.)? of the Founder herself a leper at the Leper Hospital at Hendele, a village at the mouth of the River Kelani near Colombo.*
17 × 13 cm.
Note: Monogram with date 1708.

O. 6, 13 *Dutch Presbyterian Church at Wolvendahl Colombo. The West Front.*
17 × 13 cm.
Note: Church bears date 1749. This drawing has been reproduced in *Cultureel Indië,* Vol. I, p. 34.

O. 6, 14 *South entrance Dutch Presbyterian Church, Wolvendahl. Colombo.*
17 × 13 cm.
Note: Church bears date 1749.

O. 6, 15 *Interior of Wolvendahl Church Colombo as it appeared in A.D. 1838.*
18 × 14 cm.
Note: This drawing has been reproduced in *Cultureel Indië,* Vol. I, p. 34.

O. 6, 16 *The old Dutch canal in the Fort of Colombo recently filled up.*
8 × 13 cm.

O. 6, 17 *Old Dutch street within the Fort of Colombo.*
7,5 × 13 cm.

O. 6, 18 *Entrance to the canal at Hendele. North of Colombo.*
7,5 × 12,5 cm.

O. 6, 19 *The Wellewatte Canal. South of Colombo.*
7,5 × 12,5 cm.

O. 6, 20 *Drawbridge at Grand Pass leading to the site of the old silk factory.*
8 × 14 cm.

O. 6, 21 *Lock Gate. St. Sebastian Canal. Colombo Lake.*
8,5 × 14 cm.

O. 6, 22 *Fort of Galle from the North esplanade. The Harbor & custom House on the extreme left.*
7,5 × 33 cm.

O. 6, 23 *Fort of Matura from the East. Entrance on the right.*
8 × 15 cm.

O. 6, 24 *The Star Fort or the Redoubt van Eck, Matura, on the northern bank of the river & North West of the Fort.*
8 × 15,5 cm.

O. 6, 25 *Dutch Reformed Presbyterian Church, Galle.*
17 × 12 cm.

O. 6, 26 *Interior of the Dutch Reformed Presbyterian Church at Galle.*
17 × 12 cm.

O. 6, 27 *Dutch Reformed Presbyterian Church. Matara.*
8 × 13 cm.

O. 6, 28 *Entrance to the Dutch Reformed Presbyterian Church at Matara.*
8 × 13 cm.

O. 6, 29 *Entrance to the Redoubt van Eck. The "Star Fort" of Matara.*
17 × 13 cm.
Note: Above the entrance is the date 1763 and the inscription "Redoute van Eck", monogram V.O.C.

O. 6, 30 *Gate of the Fort of Matara.*
8 × 13 cm.
Note: Gate bears date 1789.

O. 6, 31 *Inscription within the entrance to the Redoubt van Eck, or the Star Fort of Matara.*
8 × 13 cm.
Note: The inscription reads: "L. H. Lappe Capt., H. Entink Lt., Ingeneurs Exect".

O. 6, 32 *Gateway of a private dwelling house, Matara.*
8 × 13 cm.
Note: There is an inscription "De Vreede" to both sides of the gate.

O. 6, 33 *Inscription over the entrance door of a house in the Fort of Colombo.*
8 × 5,5 cm.
Note: The inscription reads: "door 't velt gewelt door regt gestelt".

O. 6, 34 *Double monogram at the residence of Commandant A. de Ly at Galle.*
8 × 5,5 cm.

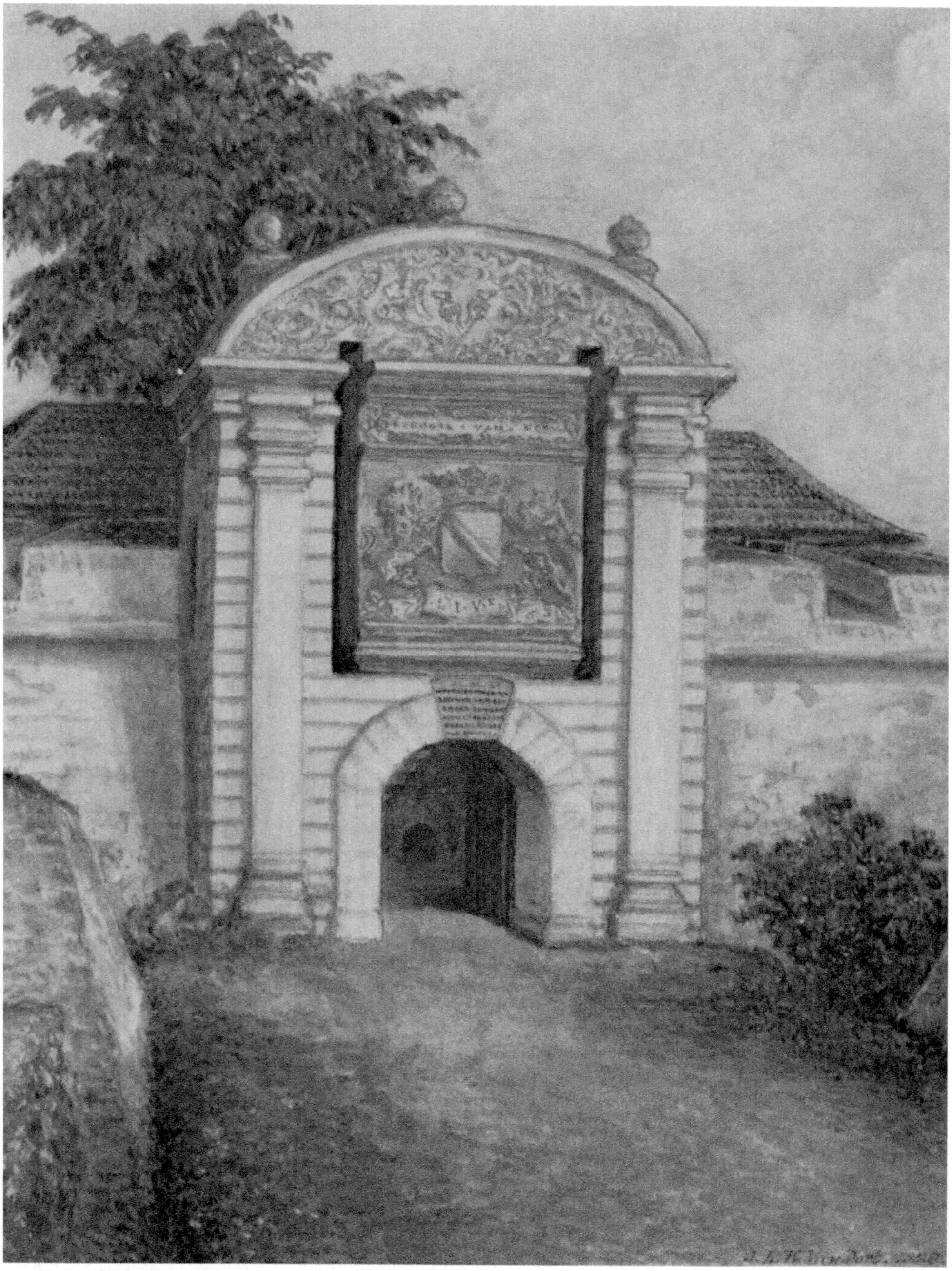

J. L. KALENBERG VAN DORT   Redoubt "Van Eck", Ceylon   O. 6, 29

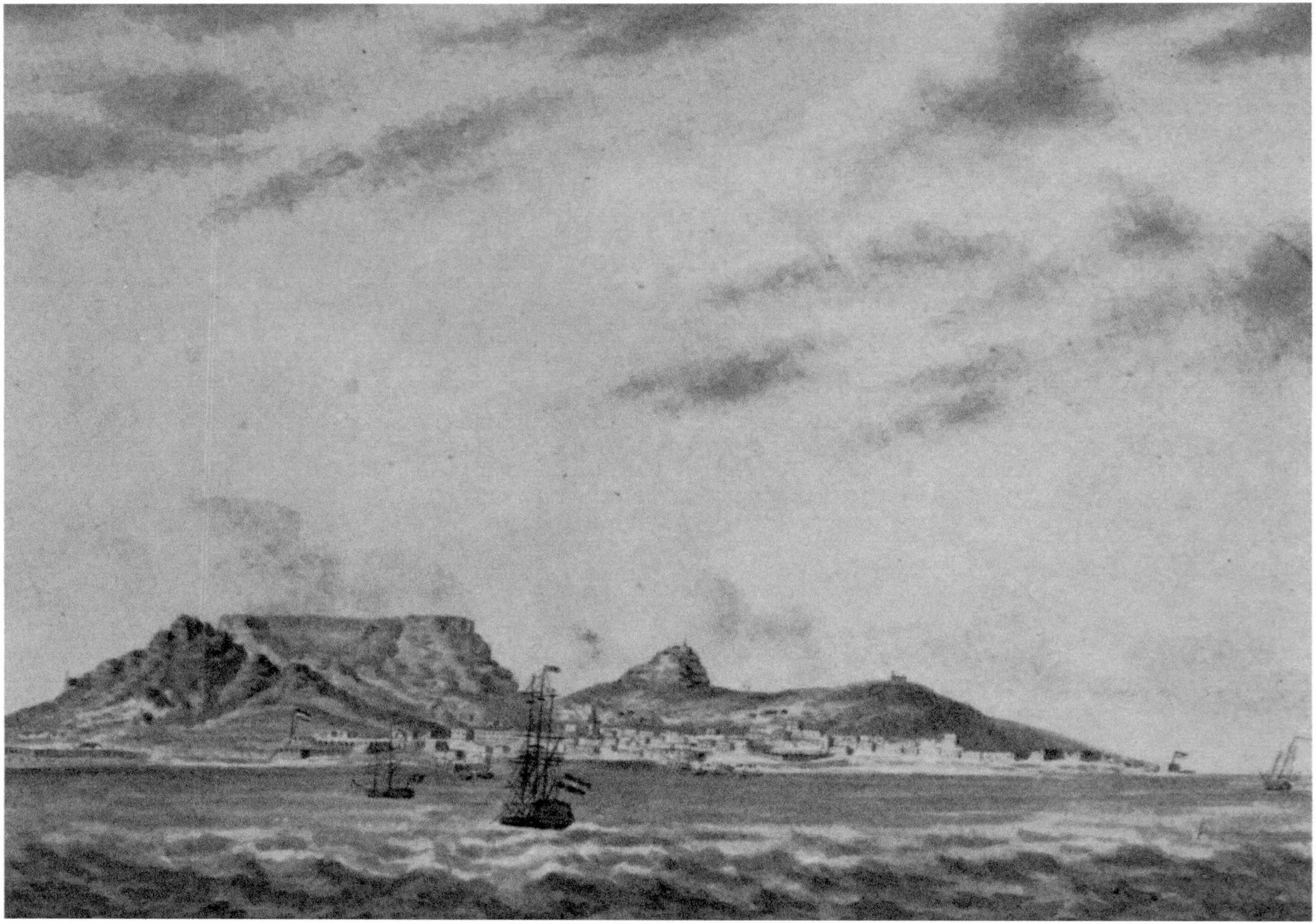

(W. B. C. PARAVICINI DI CAPELLI) Cape Town H. 447, a; *see* p. 109

O. 6, 35 *Fort of Negombo.*
8 × 6,5 cm.

O. 6, 36 *Fort of Kalutara.*
8 × 6,5 cm.

O. 6, 37 *Dutch Presbyterian Church at Wolvendahl from the inner harbor.*
8 × 6,5 cm.

O. 6, 38 *Wolvendahl Church birds-eye view form South West.*
8 × 6,5 cm.

O. 6, 39 *Batticaloa Fort.*
8 × 14,5 cm.

O. 6, 40 *Fort of Galle from the South West.*
8 × 14,5 cm.

O. 6, 41 *Fort Ostenberg & part of Trincomalie harbour.*
8 × 14,5 cm.

O. 6, 42 *Fort Frederick (with Dutch bay & the Governor's bungalow) Trincomalie.*
8 × 14,5 cm.

O. 6, 43 *The Falck Memorial in the Dutch Reformed Presbyterian Church at Wolvendahl, Colombo.*
19 × 12,5 cm.
Note: Memorial bears the inscription: "Ter gedagtenisse van den Wel Edelen Grootagtbaaren Heer Iman Wellem Falck in 't leeven Raad Ord. van Nederlands Indien Governeur en Directeur van 't eiland Ceilon en resoorts van dien geb. te Kolumbo den 25 Maart Ao. 1736 overleden te Kolumbo den 6 Febry. Ao. 1785" (In remembrance of the Right Honourable Iman Wellem Falck, late member of the council for the Dutch East-Indies, Governor and Director of the island of Ceylon and dependencies. Born in Colombo, March 25, 1736, died in Colombo on Febr. 6, 1785).

O. 6, 44 *Three wall tablets in Wolvendahl Church at Colombo. Placed in the organ loft, not represented in L. Ludovici's Lapidarium Zeylanicum.*
17 × 13 cm.
Note: Tablets bear text: 1. Van Angelbeek & Lever. 2. Van de Graaf & Van Angelbeek. 3. Unknown tablet.

O. 6, 45 *Arms, Armorial Bearings etc. of the Dutch period in Ceylon.*
17,5 × 12 cm.
Note: With text: 1. "Two edged sword blade of the Dutch period. Discovered in the Central Province having the emblem of the Vereenigde Ost Indiansche Compagnie market thereon". 2. "Engraved glass of the Dutch period in the possession of a Dutch Gentleman in Matara". a. "Glass Goblet", b. "Arms engraved in the Goblet, enlarged", c. "Masonic

Loving cup of Glass". 3. "Cameo with arms of the Kalenberg Family, reduced". 4. "Seal with arms of the Vandort Family, enlarged".

O. 6, 46 *Tablet in the Dutch Church at Matara. Omitted from the Lapidarium Zeylanicum.*
17 × 12,5 cm.
Note: Tablet bears text: "Ter gedagtenisse van het jong geboore dogterje van den Opperkoopman en dessave van Mature De E. Heer Ian Bauert, den XXII 7ber 1756 geboren en kort na dies geboorte overleden" (In remembrance of the new-born daughter of the Chief merchant and 'dessave' of Mature, Mr. Ian Bauert, born on Sept. 22, 1756, and deceased shortly after her birth).

O. 6, 47 *Two tablets in the Dutch Church at Matara, omitted from L. Ludovici's Lapidarium Zeylanicum.*
18 × 12 cm.
Note: Tablets bear inscriptions: 1. "Ter gedagtenis van de jonge juffer: Plantina Johanna de Moor, geboren te Mature den 10de November 1774, overleeden den 7de Augustus Anno Domine 1777" (In memory of Miss Plantina Johanna de Moor, born at Mature, Nov. 10, 1774, died August 7, Anno Domine 1777). 2. "Oscar L. Jonas Tranchell geb. d. 8. May A. 1793" (Oscar L. Jonas Tranchell, born May 8, 1793).

O. 6, 48 *Tablet in the Dutch Church, Matara. Incorrectly represented in the Lapid. Zeylanicum.*
17 × 12,5 cm.
Note: Tablet bears inscription: "Ter loffelicker gedagtenisse van d. Heer Mr. Frans Willem Falk in zijne leven Opperkoopman en dessave van Mature, een man voortreffelijk door zijn geboorte en deugd als andere goede hoedanigheden, weggeruckt door d. Doot die Niemand verschoont en den bloey van zijn dagen als zijnde gebrn. tot Keulen den VII Xber MVIIX en overleden tot Mature den VII Augustus Ao. MVIIXXXVII oud ses en twentig iaren, agt maanden en een dag. Hier rustende ter zaligen opstandinge" (To the glorious memory of Frans Willem Falk, Esq., late Chief merchant and 'dessave' of Mature, a man excellent by birth and virtue and other good qualities, cut off in his prime by Death who spareth no-one, having been born in Cologne, Dec. 7, 1710, and having died at Mature on Aug. 7, 1737, at the age of 26 years, 8 months and one day. He rests here until the day of resurrection).

O. 6, 49 *Tablet in the Dutch Church at Matara. Incorrectly represented in the Lapidarium Zeylanicum.*
17,5 × 12,5 cm.
Note: Tablet bears inscription: "Ter gedagtenisse van d. E.E. Heer Johann^{s} Ferdinan^{s} Crytsman Oppercoopman en Secunde van 't Gaals Commandements mitsgs. dessave deser Landen, geboren te Colombo den 17 April Anno 1709, overleden den 7xber. 1758 oud 49 iaren, 8 maanden en 20 dagn. Beati sunt in Domino Morientes" (In memory of the noble Mr. Johann^{s} Ferdinan^{s} Crytsman, Chief merchant and second-in-com-

mand in Galle as well as 'dessave' of this Country, born in Colombo on April 17, 1709, died Dec. 7, 1758, at the age of 49 years, 8 months and 20 days. Beati sunt in Domino Morientes).

Provenance: Manuscript H. 468 contains a letter by P. Daendeliker, consul for the Netherlands in Colombo, dated Sept. 7, 1889, addressed to the Count De Bylandt in London in which the water-colours and some photographs were included.

References: J. B. Siebel, "Some reminicenses of J. L. van Dort, "the Artist" ", *Journal of the Dutch Burgher Union,* Ceylon, Vol. XL, April 1950, No. 2.
F. W. Stapel, "Nederlandsche herinnerigen op Ceylon", *Cultureel Indië,* Vol. I, pp. 33—35.

HOORZEE, J. C.

Approx. 1698.

H. 47 Three water-colours and many maps are contained in manuscript H. 47 by Gerrit de Heere, Governor of Colombo "Dagelijxe aanteekeningen gedaen door . . . e.a. in de Dessavonie en de landen onder het district van Colombo, mitsgaders na Gale, Mature, Batticalo, Trincoenemale, Jaffanapatnam, Manaer, Calpatijn, enz. enz." (Diary entries by . . . and others in the 'Dessavonie' and subdistricts of the district of Colombo, as well as on the way to Gale, Mature, Batticalo, Trincoenemale, Jaffanapatnam, Manaer, Calpatan, etc. etc.).

a *'t Fort Calleture van de Z. zijde gezien* (Fort Calleture as seen from the south).
Water-colours. 39 × 32,5 cm.
Note: Inserted between p. 26 and p. 27.

b *Pagoodsberg en 't Fort Tricoenmale van de Noorderbaay aan te zien* (Hill of the Pagoda and Fort Tricoenmale as seen from the northern bay).
Water-colours. 11 × 28,5 cm.
Note: Inserted between p. 192 and p. 193.

c *Pagoodsberg en 't Fort Tricoenmale van de Zuidzijde aan te zien* (Hill of the Pagoda and Fort Tricoenmale as seen from the south).
Water-colours. 11 × 28,5 cm.
Note: Inserted between p. 192 and p. 193.

UNKNOWN ARTIST

F. 15, 1 *Perspectiv van 't Kolombose nieuwe Hospitael uit het Noorden* (Perspective view of the new Hospital in Colombo from the north).
Water-colours. 42 × 54,5 cm.
Note: With side-note: A. Oppermeesters-Wooning (Headmasters house) B. Apotheek (Dispensary) C. Apotheekers Wooning (Chemist's house)

D. Hospitael (Hospital) E. Schafbaesereij van 't Hospitael (Quarters of the hospital-cook) F. Vee-stall (Cow-shed) G. Een Gedeelte van de Wall-Muur (Part of the rampart). A floral border of fancy design (Rococo style) encircles the caption. On the reverse side of the drawing are the words: "no. 15 Perspective view of the Hospital at Colombo from the North side".

F. 15, 2 *Perspectiv van 't Kolombose nieuwe Hospitael, uit het Zuiden* (Perspective view of the new Hospital in Colombo from the south). Water-colours. 41 × 53,5 cm.
Note: Above the entrance of the hospital appears the date MDCCLXXI. On the reverse side of the drawing are the words: "no. 26. South view of the Hospital Kolombose" in different handwriting: "Afteekening van 't hospitael te Colombo" (Drawing of the hospital of Colombo).
Note: On the reverse side of both water-colours (F. 15, 1 and 2) appears an incorrect statement: "by J. Rach" (according to Mrs. J. de Loos-Haaxman possibly Van Reimer). The 13 figures in the foreground are vividly depicted by a skilful hand.

Provenance: Bought at the second-hand bookshop Israël, Amsterdam.

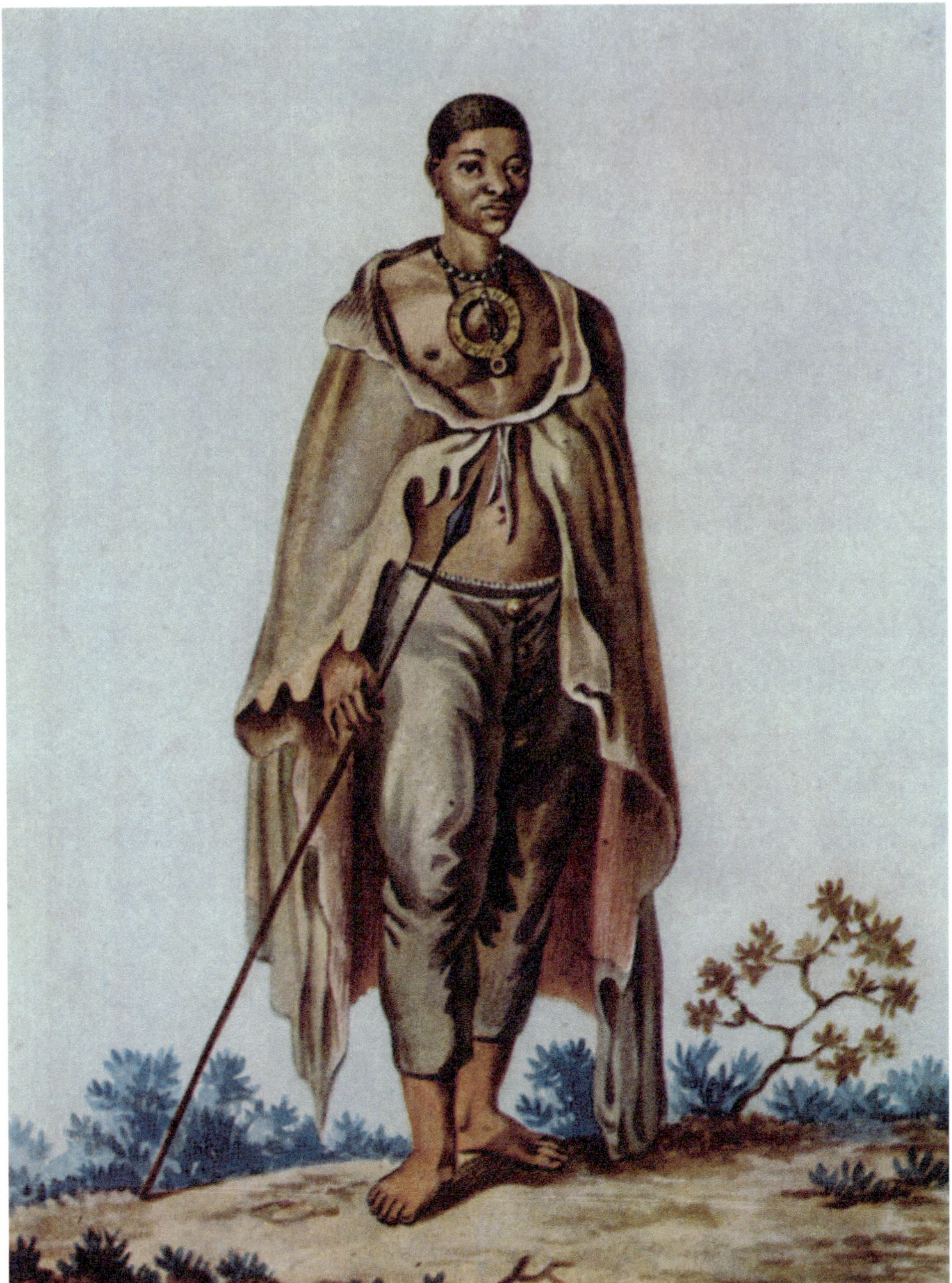

(W. B. C. Paravicini di Capelli) Andries Zwart Boy H. 477, e; *see* p. 109

SOUTH AFRICA

UNKNOWN ARTIST

H. 447 The following water-colours are included in manuscript H. 447 "Reize In de Binnenlanden van Zuid-Africa. Gedaan in den Jaare 1803 door W. B. C. Paravicini di Capelli, Kapitein Aide de Camp bij den Gouverneur van de Caap de Goede Hoop" (Journey through the Interior of South Africa in the year 1803 by W. B. C. Paravicini di Capelli, Captain and Aide de Camp of the Governor of the Cape of Good Hope). It is a copy of the original manuscript. The water-colours in it are probably copies of original drawings of the author himself. This manuscript has be edited by W. J. de Kock (Van Riebeeck-Vereniging, Kaapstad 1965), to which edition we would refer for further details and bibliography.

a *Tafelbaay en Stad de Goede Hoop* (Table bay and the Town of Good Hope).
Water-colours. Signed WP. 20 × 32 cm.

b *Overvaart aan de Breede Rivier* (Ferry across the Broad River).
Water-colours. 20 × 32 cm.

c *Groene Valley bij de Gankoma* (Green Valley near the Gaukamma).
Water-colours. 20 × 32 cm.
Note: ed. de Kock, facing p. 30.

d *Fort Frederik aan de Algoa Baay* (Fort Frederik on Algoa Bay).
Water-colours. 20 × 32 cm.
Note: ed. de Kock, facing p. 31.

e *Hottentot Andrec Zwart Boye* (A Hottentot named Andries Swartbooi).
Water-colours. 32 × 20 cm.
Note: ed. de Kock, facing p. 126.

f *Kafferkoning Gaïta* (The Kaffir king Gaïta).
Water-colours. 32 × 20 cm.
Note: ed. de Kock, facing p. 126.

g *Graaf Reinet met de Compas Berg* (Graaf Reinet and the Compass Hill).
Water-colours. 20 × 32 cm.

h *Theebos Bergen bij Graaf Reinet* (Theebos mountains near Graaf Reinet).
Water-colours. 20 × 32 cm.

i *Afrikaansche Leeuw* (African Lion).
Water-colours. 32 × 20 cm.

j *Oranje Rivier* (Orange River).
Water-colours. 20 × 32 cm.

k *Gnou* (Gnu).
Water-colours. 32 × 20 cm.

l *Campement aan de Wijnbergen* (Encampment near the Wine Mountains).
Water-colours. 20 × 32 cm.

m *Afrikaansche Leeuw* (African Lion).
Water-colours. 32 × 20 cm.

Provenance: Bought in approx. 1915 by J. W. IJzerman from Brill's in Leiden and presented to the Institute on Sept. 22, 1917.

JAPAN

UNKNOWN ARTIST

F. 14, 23—26 *Nagasaki colour-prints.*

F. 14, 23 *Portrait of a foreigner.*
Coloured woodcut, probably Yamatoya print, large panel. 44,5 × 16 cm.
Note: According to the Japanese caption, this portrait represents a "kapitan" or Dutch Director. Probably, this was either Hendrik Doeff, who was Director in Nagasaki on the isle of Deshima from 1804—1817, or his successor Jan Cock Blomhoff who was there from 1818—1824. This print has been reproduced by Mody, plate 117, fig. 1.

F. 14, 24 *Dutch lady with a parrot.*
Coloured woodcut with blind tooling, Yamatoya print, large panel. 44,5 × 16 cm.
Note: The lady represented is possibly Mrs. Titia Cock Blomhoff-Bergsma, aged 31, who was on the isle of Deshima for a short time in 1818. According to Dr. C. Ouwehand, curator of the Japanese section of the National Ethnographic Museum in Leiden, she might be Mimi de Villeneuve, who was in Japan in 1829. Reproduced by Mody, plate 92.

F. 14, 25 *Orandajin danjo no zu* (portrait of a Dutch man and woman).
Coloured woodcut (Yamatoya print ?). 31 × 21,5 cm.
Note: Represented are probably Mrs. Cock Blomhoff with her husband and child. Mody, plate 93, fig. 2.

F. 14, 26 *Hollandsche Schip* (Dutch vessel).
Hand-coloured woodcut, Bunkindo print. 31 × 21,5 cm.
Note: A Dutch vessel under full sail mounting 14 guns. Mody, plate 63, fig. 2.

References: N. H. N. Mody, *A Collection of Nagasaki Colour-prints and Paintings,* London/Kobe 1939, Vol. I.
C. R. Boxer, *Jan Compagnie in Japan, 1600—1850,* The Hague 1950.

SCULPTURES

LACOMBLÉ, Eugène

Born Sept. 5, 1828, Brussels. Died Nov. 17, 1909, Arnhem.
He was trained at the Brussels Academy of Arts and later by D. Cogniet at Paris. In 1855 he lived in The Hague and in 1865 he was appointed professor at the Technical College at Delft where he taught modelling until 1900. One of his pupils was Charles van Wijk (mentioned below). His works include: Monument to General Pel in Atjeh; Monument to Bultzing Löwe in Surabaya (v. Von Faber, "*Oud Soerabaja*", p. 124).

B. 1 *Medallion-portrait in laurel wreath, of C. F. Winter Sr.*
Plaster cast. Approx. 1864. 110 × 65 cm.
Note: C. F. Winter Sr. (1799—1859) was a well-known scholar of Javanese. The portrait was intended to be used for the monument in the grounds of the governor's house in Surakarta. The monument was unveiled Febr. 19, 1867, and has probably been destroyed in recent times.

Provenance: Presented by the sculptor, Sept. 15, 1900.

References: D. H., "Carel Frederik Winter", *Tijdschrift voor Nederlandsch-Indië,* 3rd series, Vol. I, 1867, pp. 523—531 (with illustration).

B. 2 *Design for a statue of the Governor-General Jan Pietersz. Coen.*
Plaster cast. Approx. 1875. Height 75 cm.
Note: Probably this design was made as an entry in a competition, either for the statue on Waterloo Square in Weltevreden, unveiled Sept. 4, 1876, or for that in Hoorn, unveiled May 30, 1893. As this figure points downward, like the finished statue in Weltevreden, it is most likely a design for the latter.

Provenance: Presented by the sculptor, Sept. 15, 1900.

SCHULTSZ, J. C.

Sculptor in Amsterdam.

B. 3 *Mr. C. Th. van Deventer. De vriend van Indië* (Mr. C. Th. van Deventer. Friend of the Indies).
Bronze plaque. 95 × 65 cm.

Provenance: Indonesian admirers of van Deventer presented this plaque to the Indisch Genootschap (Indian Society) on May 26, 1917. At the request of this Society it was installed in the Institute's building.

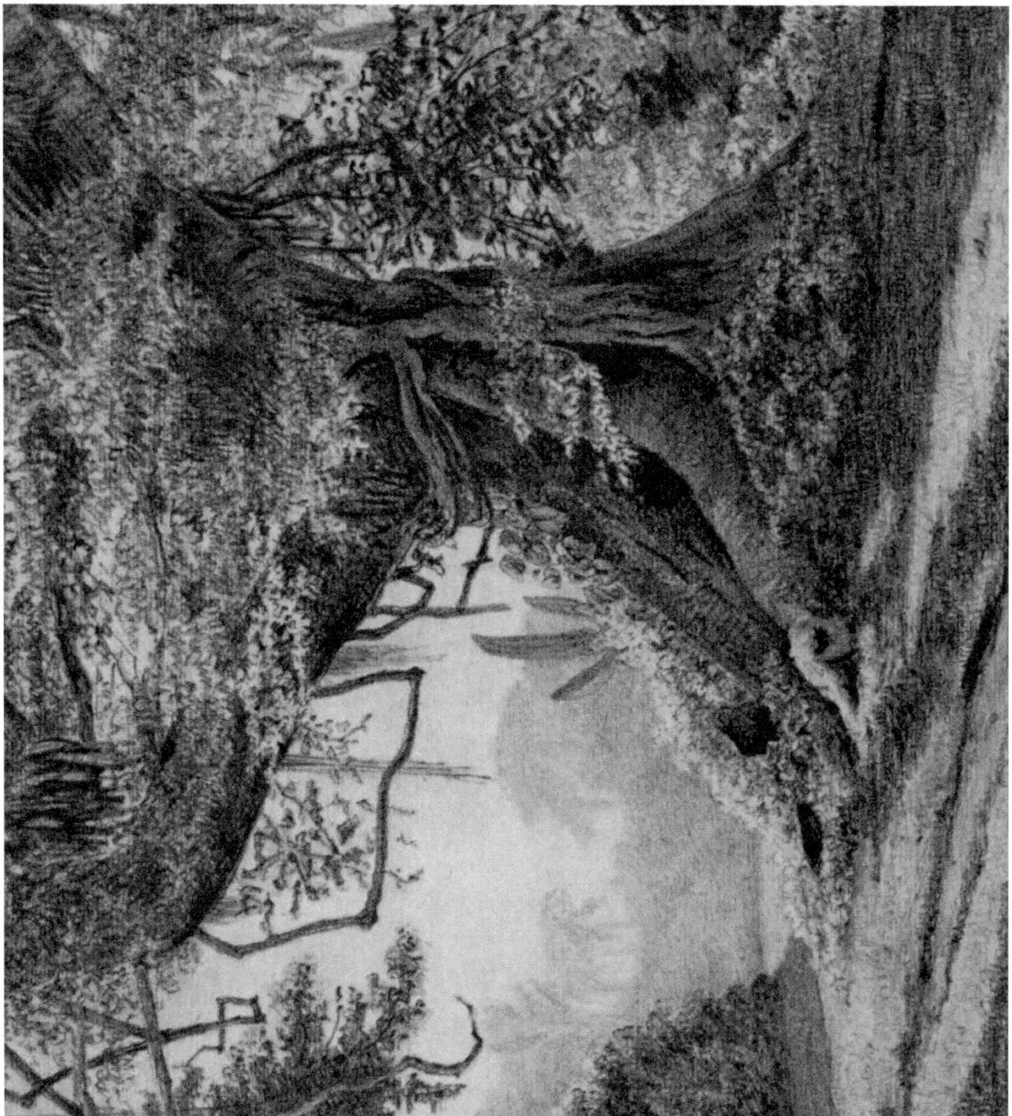

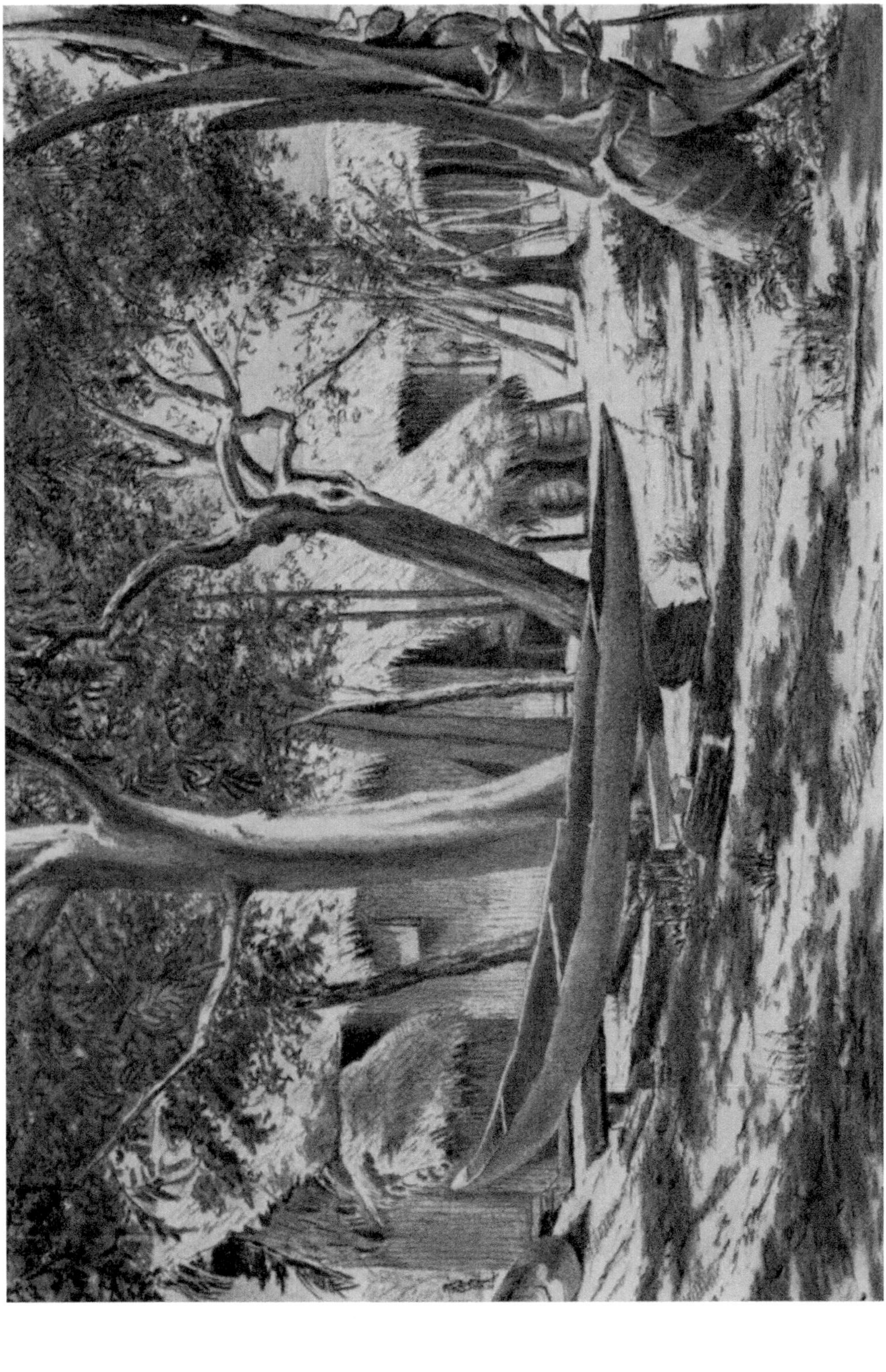

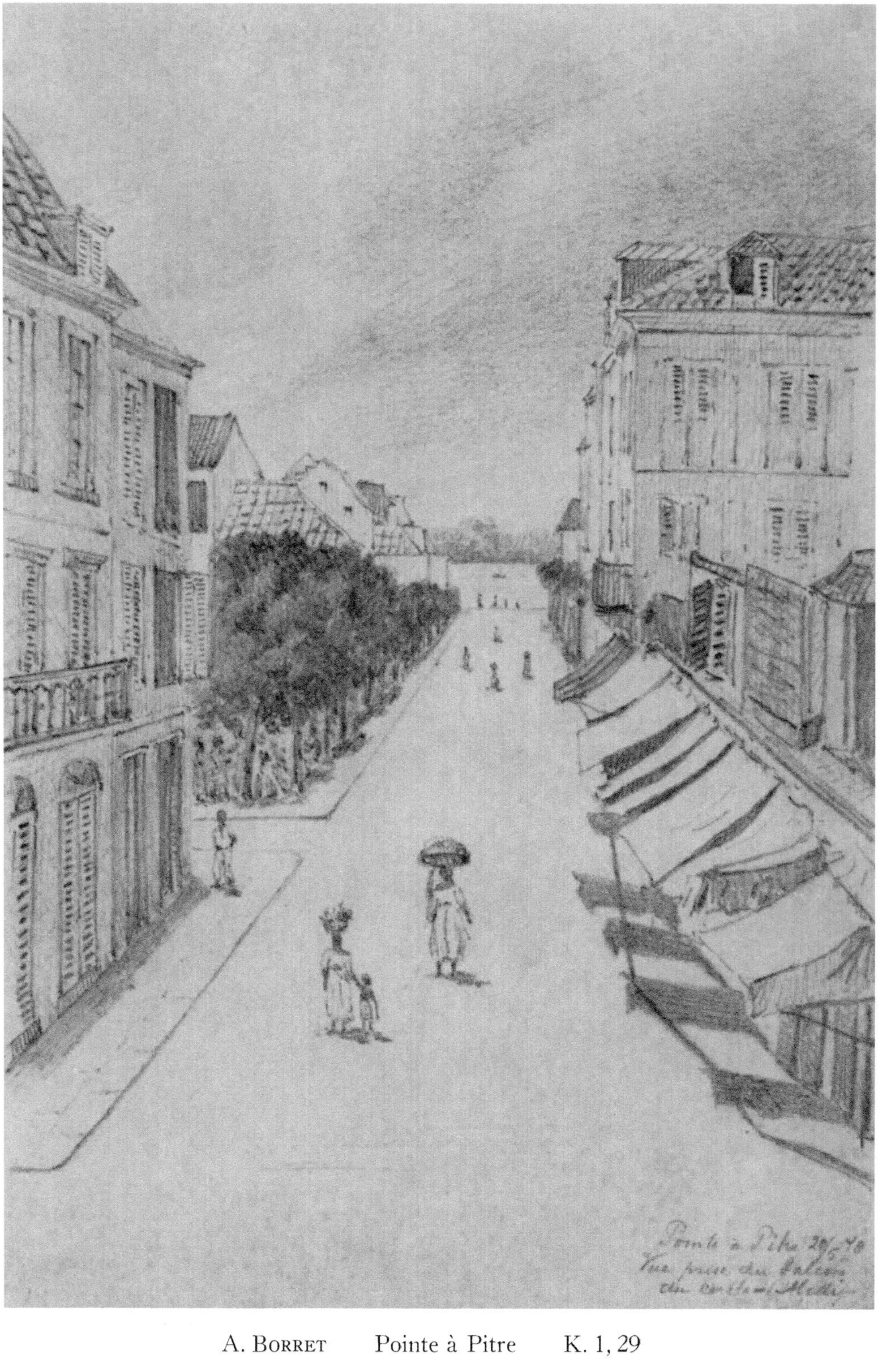

A. BORRET Pointe à Pitre K. 1, 29

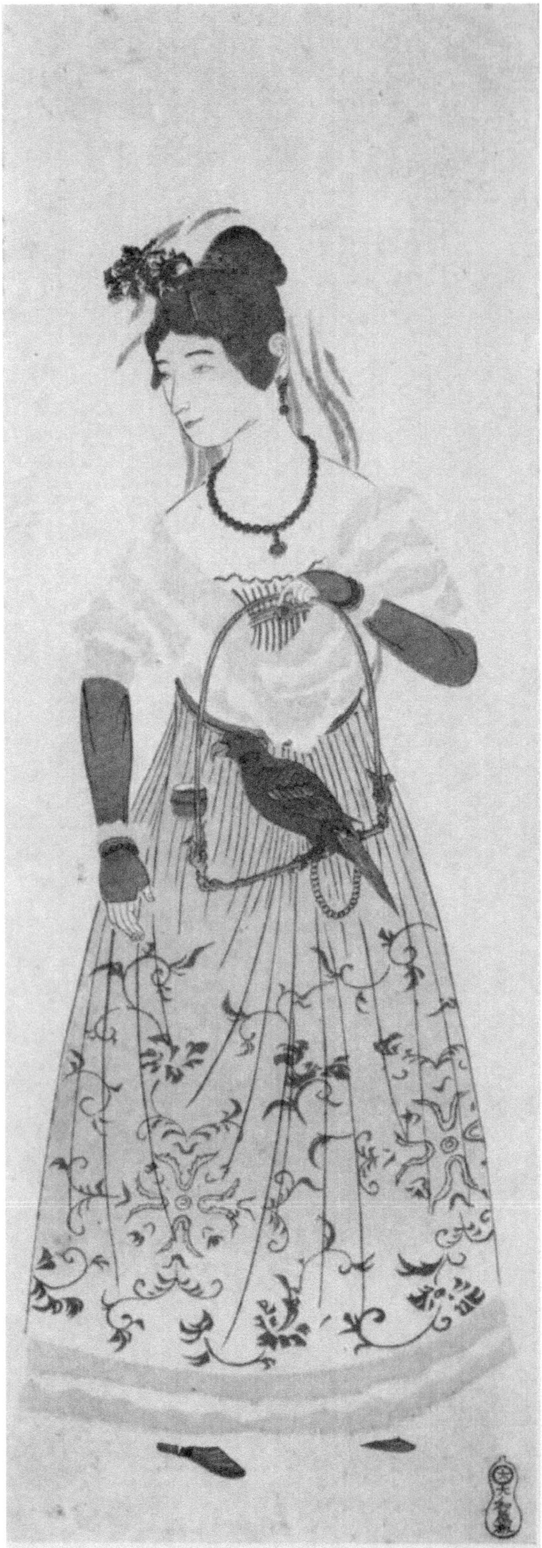

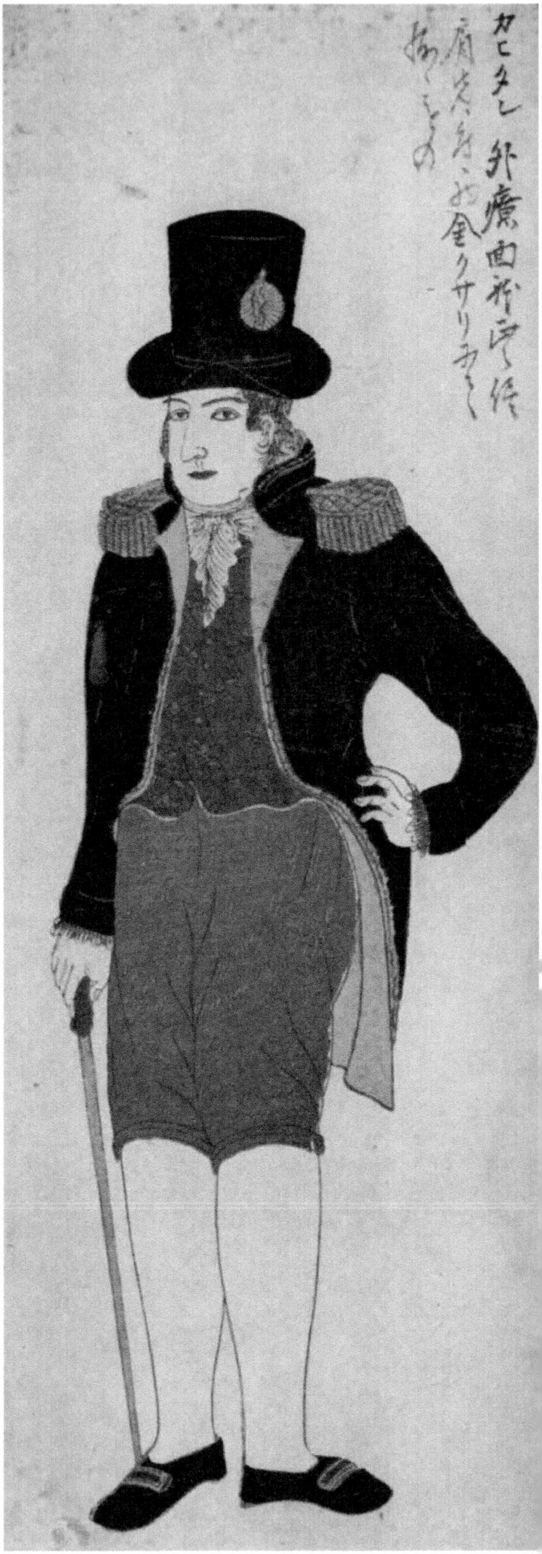

Nagasaki colour-prints

Lady with a parrot F. 14, 24

Portrait of a foreigner F. 14, 23

References: E. A. Kayadoe, "Bij de plaquette van C. Th. van Deventer", *Nederlandsch Indië Oud en Nieuw* 1917—1918, pp. 73—74 (with illustration).
Minutes of the Royal Institute June 16, 1917, *B.K.I.* LXXIII, 1917, p. XXV.

WIJK, Charles van

Born 1875. Died 1917.
Pupil of Lacomblé, less academic than his master, more an impressionist. His works include monuments (Major Thomson in the Hague; prof. Van 't Hoff and Willem Schuurman in Rotterdam) and portrait-heads (Bilderdijk in the Nat. Museum, Amsterdam).

B. 4 *Bust of Prof. H. Kern.*
Marble on pedestal. 72 × 50 × 35 cm.
Note: The bust was offered to professor H. Kern (1833—1917) by a committee on the occasion of his 80th birthday. On behalf of this committee professor J. S. Speyer officially entrusted the bust to the care of the board of directors of the Royal Institute during its meeting on Oct. 18, 1913. A plaster cast of this bust is in the Kern Institute in Leiden.

INDONESIAN ARTISTS

DJAJENG SUDIRDJO, Radèn Bekel

GF. 2 *Gambar-djikaloe Sri Padoeka Sulthan Djoegjakarta mijos di Residentnan tanggal 1 Januari dan 31 Augustus. Terbikin olih Bekel penakawan Midji die Kraton Djoegjakarta* (Picture representing the visit of the Sultan of Djogjakarta to the Resident's house on the first of Januari and on 31st August. Made by a servant of the Kraton in Djogjakarta).
Water-colours and gold. Signed R. B. Djajeng Soedhirdja. Dated 5/10, 1904. 25 × 2700 cm.
Note: Originally a scroll consisting of 30 strips of 25 × 90 cm, pasted together, in a box with caption "Sultan van Djokja in Statie" (Sultan of Djogjakarta in State). Recently framed in 17 gilt frames.

Provenance: Bought in Dec. 1958, from E. J. R. Poublon, formerly employed on a sugar plantation near Djogjakarta; the original owner was "Pa'tje Versteegh" (Daddy Versteegh), i.e. W. Versteegh of the Bantool sugar refinery near Djogjakarta, who died in Bandung about 1955.

Reference: J. Terwen-de Loos, "De gouden koets van de Sultan", *B.K.I.* CXXIII, 1967, pp. 366—372.

GF. 3 *Gambar djikaloe Srp. (Sri Padoeka) Kangdjeng-Sultan tedak di Resideennan tanggal 1 Januari en 31 Augustus. Terbikin oleh poenokawan midji . . .* (Picture representing the visit of the Sultan to the Resident's house. Made by a special servant of the Royal house).
Water-colours and gold on linen. Four scrolls each 23 × 550 cm.
Note: Drawing enclosed in box with title "Sultan van Djokja in Statie" (Sultan of Djogjakarta in State). Reduced copy based on GF. 2.

PASEK, I Dewa Made

Balinese artist in Batuan, South Bali.

S. 19 *Bhimaswarga.*
Tempera on cotton cloth, recently reframed. Signed in pencil on reverse side: Dew Mad Pasek, Batuan. 60 × 104 cm.
Note: Representations of torments of hell, placed in a Balinese landscape. Bhima is represented sitting in a tree. The artist is probably not to be identified with I Dewa Kompiang Pasek, in Batuan, born 1924; cf. *Hedendaagse Kunst van Bali,* Utrecht 1962, p. 92.

Provenance: Bought from Van der Peet's antiquary, Amsterdam, 1964.

RUDIN, I Ktut

Balinese artist in Sanur, South Bali.
F. 17, 1—3 *Three pictures of Balinese woman dancers.*
Water-colours and ink on paper. 31,5 × 24 cm.
Note: Gestures and postures based on sketches of movements of the *legong* dance by Miguel Covarrubias in his *Island of Bali,* New York 1937, between pp. 226 and 227. Cf. Th. P. Galestin's notes in *Hedendaagse Kunst van Bali,* Utrecht 1962, pp. 73—75.

Provenance: Bought from Van der Peet's antiquary, June 1967.

SALEH SARIEF BASTAMAN, Raden

Born 1814, Semarang. Died 1880, Buitenzorg.
Pupil of the Belgian painter A. A. J. Payen, who accompanied C. G. C. Reinwardt during his travels. After training in Europe, Raden Saleh made many romantic pictures of Indonesian landscapes and several scenes with animals. He also visited Algeria. The pencil-drawing in the collection of the Royal Institute is a preliminary study of one of his many portraits in oil-paints.

1—46—66 *Pangeran Sarif Hamid Alkaddrie.*
Pencil. Signed R.S.f. Dated 1853. 24 × 19 cm.

SASTROSENDJOJO, M.

Native teacher at Bantul.

L. 922, 227, No. 14 *Map of Pasar Gede (Kuta Gede).*
Water-colours. 34 × 72,5 cm.
Note: This map is preserved in the collection of maps. It is a prospective projection of the royal tombs with comments in Dutch. At the bottom left: the "blue stone" under a penthouse roof (probably the royal seat).

Provenance: From the collection of W. F. Engelbert van Bevervoorde.

SOERIODIKOESOEMO, R. A.

In manuscript H. 468 (Lapidarium of the Dutch Indies compiled at the initiative of the Royal Institute, 1888) there are two water-colours of a gate called "Labang Buta" (Giant Gate) in Bangkalan, Madura.

a *Labang Boeta. Palitan dari sabela barat* (Labang Buta as seen from the west).
Water-colours. Signed R. A. Soeriodikoesoemo. 34 × 27 cm.

b *Labang Boeta. Paliatan dari sabela timoer* (Labang Buta as seen from the east).
Water-colours. Signed R. A. Soeriodikoesoemo. 34 × 27 cm.

TRENGONO, Raden Mas

Son of Ario Tjondro Negoro, regent of Demak, regency of Semarang.

K. 28, 14 *Types of umbrellas.*
Water-colours and gold. Signed Raden Mas Trengono. Dated Samarang September 1864. 21 × 27 cm.
Note: There are 28 drawings of umbrellas, and a French caption stating the name of the artist.

K. 28, 15 *Types of umbrellas.*
Identical with K. 28, 14, but dated December 1864 and with the caption in German.

UNKNOWN ARTISTS

K. 4, 1—5 *Five foolscap sheets with drawings of 27 native weapons.*
Water-colours. Dated 1886. 42 × 34 cm.

K. 4, 1 (1) *Kris (gabèlan).*

K. 4, 1 (2) *Kris (Javaansche)* (Kris (Javanese)).

K. 4, 2 (3) *Kris (Balinesche)* (Kris (Balinese)).

K. 4, 2 (4) *Lemmet van een kris (pĕndòwò)* (Blade of a kris (pĕndòwò)).

K. 4, 2 (5) *Lemmet van een kris (Sepoehal)* (Blade of a kris (Sepuhal)).

K. 4, 2 (6) *Lemmet van een kris (djalak)* (Blade of a kris (djalak)).

K. 4, 3 (7) *Piek (toembak)* (Pike (tumbak)).

K. 4, 3 (8) *Piek (toembak)* (Pike (tumbak)).

K. 4, 3 (9) *Piek (Sangkoeh), met een blaasroer tot stok* (Pike (Sangkuh), the shaft consisting of a blow-pipe).

K. 4, 3 (10) *Lemmet, behoorende bij no. (K. 4, 3) 9* (Blade belonging to No. 9).

K. 4, 3 (13) *Pangoh, tot het vervaardigen van kleine houten voorwerpen* (Pangoh, used for the construction of small wooden objects).

K. 4, 3 (14) *Grasmes (Arit)* (Sickle (Arit)).

K. 4, 4 (11) *Lemmet, behoorende bij no. (K. 4, 3) 7* (Blade, belonging to No. 7).

K. 4, 4 (12) *Lemmet, behoorende bij no. (K. 4, 3) 8* (Blade, belonging to No. 8).

K. 4, 4 (15) *Ponjaard (tjoendrik)* (Poniard (tjundrik)).

K. 4, 4 (16) *Ponjaard (djambea)* (Poniard (djambea)).

K. 4, 4 (17) *Steekmes (wedoeng soedoek, groot soort)* (Pointed dagger (wedung suduk, large type)).

K. 4, 4 (18) *Hakmes, groot soort (wedoeng dòwò)* (Matchet, large type (wedung dòwò)).

K. 4, 4 (19) *Hakmes, klein soort (wedoeng tjĕndĕk)* (Matchet, small type (wedung tjĕndĕk)).

K. 4, 4 (20) *Hakmes (koedie)* (Matchet (kudi)).

K. 4, 4 (21) *Hakmes (tjaloek)* (Matchet (tjaluk)).

K. 4, 5 (22) *Werpspies (tjĕlèngkrèng)* (Javelin (tjĕlèngkrèng)).

K. 4, 5 (23) *Werpspies (Nĕnggòlò)* (Javelin (Nĕnggòlò)).

K. 4, 5 (24) *Werpspies (Limpoeng)* (Javelin (Limpung)).

K. 4, 5 (25) *Hakmes (Lamĕng)* (Matchet (Lamĕng)).

K. 4, 5 (26) *Hakmes (Tjĕloerit)* (Matchet (Tjĕlurit)).

K. 4, 5 (27) *Hakmes (Larkang)* (Matchet (Larkang)).

★

K. 4, 6—28 *Teekeningen van Javaansche Krissen en Lansen* (Drawings of Javanese krisses and lances).
Indian ink. 23 foolscap sheets. 34 × 42 cm.
Note: This collection contains 115 drawings with Javanese and Dutch captions stating the Javanese name of each type of kris.

★

K. 4, 29 *Lemang-Gandring en Sengkelat* (Lemang-Gandring and Sengkelat).
Water-colours. 11 foolscap sheets. 27 × 21 cm.
Note: This is an account, richly illustrated with small pictures of krisses and lances.

K. 4, 30—47 *18 quires (originally 39) with illustrations of Javanese and other weapons, some of which are accompanied by illustrated comments.*
Water-colours. Foolscap sheets. 34 × 21 cm.
Note: A list of contents is used by way of cover. The Nos. 1, 3, 6, 8, 10, 13, 15, 17, 19—21, 23, 26, 28, 30, 32, 34—38 mentioned in this list are missing from the collection.

K. 4, 30 *Nogo-Simo.*
Note: Kris with picture of a tiger.

K. 4, 31 (1) *Kris opgetooid* (Decorated kris).

K. 4, 31 (2) *De kris of Wesi-adji met handvat* (The kris or Wesi-adji and its hilt).

K. 4, 31 (3) *De houten Rongko* (The wooden Rongko).

K. 4, 31 (4) *De koperen Pendoh* (The brass Pendoh).

K. 4, 31 (5) *Wesi-adji met Gondjo* (Wesi-adji with Gondjo).

K. 4, 31 (6) *Peksi.*

K. 4, 31 (7) *Mendoh.*

K. 4, 31 (8) *Oekiran.*

K. 4, 32 *9 toembak's (lances) etc.*

K. 4, 33 *Sempono Boengkem.*

K. 4, 34 *Wesi-adji Singkir Geni.*
Note: With pictures of a fire and a rice-barn.

K. 4, 35 *Rongko's en Oekiran's* (Rongkos and Ukirans).

K. 4, 36 *Botjah angon and three other krisses.*

K. 4, 37 *Wesi-adji Brodjol Loog Senoengal.*

K. 4, 38 *Variaties [van krissen]* (Variations [of krisses]).

K. 4, 39 *Kodi-Trantjang.*

K. 4, 40 *Pamors* (Various kinds of Pamors).

K. 4, 41 *Panah klajan Gendewoh. Pijl en boog* (Panah klajan Gendewoh. Bow and arrow).

K. 4, 42 *Vrouwendracht* (Type of kris carried by women).

K. 4, 43 *Pedang Loewoh.*

K. 4, 44 *Toelichtingen. Hoe een Inlander een kris koopt* (Explanations. Native way of buying a kris).

K. 4, 45 *Klewang en Rentjong van Atjeh* (Klewang and Rentjong of Atjeh).

K. 4, 46 *Wapens van Bandjarmasin* (Weapons from Banjarmasin).

K. 4, 47 *Boedek Bésoe* (Budek Besu).

★

K. 28, 1 *Statie Pajong's van Djokjakarta* (State umbrellas from Jogjakarta).
Water-colours and gold. 32 × 42 cm.
Note: 2 double sheets with 2 x 20 cut-out figures of "Pajong's van de Vorstelijke familie" (Umbrellas of the Royal family) and "Pajong's van Sultan's ambtenaren" (Umbrellas of the Sultan's officials), with Dutch captions.

K. 28, 2 *Statie-mutsen door de vorstelijke familie en Sultan's Ambtenaren gedragen* (State caps worn by the royal family and the officials of the Sultan).
Water-colours and gold. 32 × 47 cm.
Note: 1 double sheet with 7 cut-out figures, with Dutch captions.

K. 28, 3 *Broekstoffen die uitsluitend door de Vorstelijke familie en ambtenaren mogen worden gedragen* (Trouser materials which may be

worn exclusively by the Royal family and officials).
Water-colours. 29 × 19,5 cm.
Note: 3 drawings pasted on 1 sheet, with Dutch captions.

K. 28, 4 *Stoffen van Kains die uitsluitend door de vorstelijke familie en ambtenaren gedragen mogen worden* (Kain materials which may be worn exclusively by the Royal family and officials).
Water-colours.
Note: 11 drawings pasted on a number of sheets arranged in a quire, with Dutch captions.

K. 28, 4 (1) *Hoek.*
11,5 × 18,5 cm.

K. 28, 4 (2) *Parang roesak* (Parang rusak).
17 × 17,5 cm.

K. 28, 4 (3) *Sĕmèn agĕng sawat garoeda* (Sĕmèn agĕng sawat garuda).
15 × 18,5 cm.

K. 28, 4 (4) *Oedan riris* (Udan riris).
9 × 12 cm.

K. 28, 4 (5) *Roedjak sèntė* (Rudjak sèntė).
11,5 × 18,5 cm.

K. 28, 4 (6) *Kawoeng* (Kawung).
9,5 × 15 cm.

K. 28, 4 (7) *Sĕmèn agĕng.*
14 × 17 cm.

K. 28, 4 (8) *Sĕmèn alit.*
19,5 × 16 cm.

K. 28, 4 (9) *Slobok.*
16,5 × 18,5 cm.

K. 28, 4 (10) *Tjlatjap Kemitir.*
16 × 18,5 cm.

K. 28, 4 (11) *Sĕmar mèsĕm.*
16,5 × 19 cm.

K. 28, 5 *De Goenoengan's dat zijn spijzen in reusachtigen en sierlijken vorm, die op Garebegdagen voor het Offermaal worden opgedragen* (The Gunungans ("mountains"), i.e. food served in huge decorative pyramids, only used on Garebeg (feast) days for the sacrificial meal).
Water-colours. 24 × 41 cm.
Note: 10 drawings on one double sheet, with captions and comments in Dutch. 1. *G. Koetoek.* 2. *Pitjisan.* 3. *G. Lanang.* 4. *G. Wadon.* 5. *G. Darat.* 6. *G. Pawoean.* 7. *Gepak.* 8. *Songom.* 9. *Tebok-angkring.* 10. *Mand met rijst* (Basket with rice).

K. 28,6 *Schetskaart van het Strand bezuiden Jogjakarta* (Sketch-map of the beach south of Jogjakarta).
Water-colours. 26,5 × 39,5 cm.

K. 28, 7 *Doorsnede van de rots te Rongkob* (Profile of the rock at Rongkop).
Ink. 22 × 21 cm.

K. 28, 8 *De Goenoengan Lanang, een der spijsbergen, die op een Garebegdag door 24 man wordt opgedragen* (Gunungan Lanang, one of the pyramid-shaped piles of food which is carried by 24 men on a *garebeg* (feast) day).
Water-colours. 20,5 × 36 cm.

K. 28, 9 *Zoutbereiding* (Salt-extraction).
Water-colours. 17,5 × 12 cm.
Note: There is a Dutch note explaining how three kinds of salt are extracted from boiled sea-water.

K. 28, 10 *Salt-extraction.*
Water-colours. 35 × 27 cm.

Provenance: K. 28, 1—10 from the collection of W. F. Engelbert van Bevervoorde.

★

K. 28, 11 *Pajoengs van ambtenaren en officieren onder Mangko Negårå* (Umbrellas of officials and officers of Mangkunegara).
Water-colours and gold. Dated 1869. 41,5 × 33 cm.
Note: There are 66 drawings of umbrellas with Javanese captions.

K. 28, 12 *Pajoengs in gebruik bij de leden der keizerlijke family en bij de ambtenaren binnen het gebied van den Soesoehoenan van Soerakarta* (Umbrellas used by the members of the imperial family and the officials within the realm of the Susuhunan of Surakarta).
Water-colours and gold. Dated 1869. 41,5 × 33 cm.
Note: There are 63 drawings of umbrellas with Javanese captions.

K. 28, 13 *Jogjakarta umbrellas.*
Water-colours and gold. Stitched quire of 15 sheets. 21,5 × 27 cm.
Note: 26 drawings of folded umbrellas in their stands, with Javanese and Malay captions stating for each umbrella the ranks of those who are entitled to use it.

Provenance: K. 28, 11—13 from the collection of J. P. Zoetelief, who was resident of Surakarta in 1867—1869.

★

K. 28, 16 *Sultans-statietroepen* (Sultan's ceremonial troops).
Water-colours. Folded sheet of 34,5 × 40,5 cm.
Note: Fifteen pictures of standard-bearers of different military corps of

the Sultan of Jogjakarta in their ceremonial attire, with Dutch explanations.

Provenance: Probably from the collection of W. F. Engelbert van Beevervoorde.

★

K. 31 *Statietroepen Sultan* (Sultan's ceremonial troops).
Water-colours. Bound volume of 33,5 × 22,5 cm.
Note: There are 41 pictures of Javanese soldiers and standard-bearers of different corps of the Sultan of Jogjakarta with Javanese introduction and captions.

★

K. 29 *Batik patterns.*
Water-colours on tracing-cloth. Mostly 33 × 23,5 cm; some 33 × 47 cm.
Note: In leather cover decorated with coloured ornamental design. The majority of captions in Roman script, some in Javanese script.

K. 29, 1 *Kawoeng (plentong).*

K. 29, 2 *Kawoeng poetri.*

K. 29, 3 *Kawoeng pitjis.*

K. 29, 4 *Parang Roesak (Barong).*

K. 29, 5 *Parang Roesak (Kutik).*

K. 29, 6 *Pari kèsit.*

K. 29, 7 *Pamprim (latar poetih).*

K. 29, 8 *Parang Koesoemo.*

K. 29, 9 *Parang Tjentoeng.*

K. 29, 10 *Oedan Riris (latar poetih).*

K. 29, 11 *Oedan Riris (latar ireng) or Roedjak Sénté.*

K. 29, 12 *Taroentoem.*

K. 29, 13 *Grompol tek.*

K. 29, 14 *Bantji Kaboet.*

K. 29, 15 *Poetjang Remenggo.*

K. 29, 16 *Sembagèn gambir sekèrèk or Tjeplok Bentaroem.*

K. 29, 17 *Sekar or Kembang Randoe.*

K. 29, 18 *Sembagen Hook.*

K. 29, 19 *Hook.*

K. 29, 20 *Songgo langit.*

K. 29, 21 *Gabah-sinawoer.*

K. 29, 22 *Sèmĕn Romo.*

K. 29, 23 *Tjoewiri.*
Note: Double sheet.

K. 29, 24 *Sekar or Kembang Polo.*

K. 29, 25 *Rengganis or Kawoeng Bendi.*

K. 29, 26 *Rengganis or Kawoeng Bendi.*

K. 29, 27 *Wora-wari Roempak.*

K. 29, 28 *Tambal Kanoman.*

K. 29, 29 *Tjeplok Tjepoko.*

K. 29, 30 *Merak-ngigel.*

K. 29, 31 *Pandelegang.*

K. 29, 32 *Slobak (?) Baris Koendoer (?).*

★

F. 14, 28 *De Koningin van het Zuiden verschijnt aan den als kluizenaar op een rots in zee vertoevende jongeling Prawitto en legt hem uit, hoe hij koning van Ngoban kan worden* (The Queen of the South appearing to the youth Prawito who has retired as a hermit to a rock in the sea, and explaining to him how he may become king of Ngoban).
Water-colours. 16,5 × 31,5 cm.
Note: With caption "No 2426" (crossed out) "Uit de legende van Prawitto" (From the legend of Prawito). Probably the work of a Javanee artist, who intended to depict Njai Lara Kidul, followed by the host of the sea, appearing before Prawitå, who is sitting on a small island in the ocean. In the upper left corner there is a pencil note saying: "Hoofdstuk 11" (Chapter 11) (or 17?).

F. 14, 29 *Vorst Mangkoe Rat II doorsteekt met zijn kris, genaamd "de eerwaarde Blabor" den opstandeling Troenådjåjå, dien hij met zijne twee vrouwen Klitïng Koening en Klitïng Woengoe, zusters van den vorst voor zich had laten komen* (Prince Mangku Rat II stabs his kris, called "the reverend Blabor" (Belabar), the rebel Trunadjaja, whom he had ordered to appear before him together with his two wives Kliting Kuning and Kliting Wungu, sisters of the prince).
Water-colours and gold. 27,5 × 38 cm.
Note: Drawing made by a Javanese artist. This story appears in the *Babad Tanah Djawi*; translated by W. L. Olthof, 's-Gravenhage 1941, pp. 202—203.

Provenance: F. 14, 28 and F. 14, 29 probably from the collection of W. F. Engelbert van Bevervoorde.

F. 14, 27 *Soenan Mas verlustigt zich in de aanschouwing der door hem bevolen straffen en wreedheden. 1703* (Sunan Mas gloating over the punishments and cruelties being inflicted on his orders. 1703).
Water-colours and gold. 35 × 42,5 cm.
Note: This title on reverse side of the drawing. Subscription of the drawing: "Raden Ajoe Lembah, eene der vrouwen van vorst Soenan Mas, dochter van vorstens oom Prins Poeger en Raden Soekrå, met wien zij een verboden minnehandel hield, op last en ten aanschouwe van genoemden vorst beiden geworgd, eerstgenoemde door een harer broeders (zoon van Prins Poeger). Hare talrijke vrouwelijke bedienden naakt in het tijgerhok geworpen. Een andere zoon van Prins Poeger, genaamd Raden Sotjo Koesoemo, die opstand trachtte te verwekken op de aloen-aloen in eene kooi ten toon gesteld. Dit geschiedde in het jaar 1625 der Javaansche en in het jaar 1703 der Europeesche jaartelling" (Raden Aju Lembah, one of King Sunan Mas's wives, daughter of the king's uncle, Prince Puger, and Raden Sukra, with whom she had an illicit love affair are being strangled on the orders of said king, Raden Aju by one of her brothers (a son of Prince Puger). Her numerous maid servants are being thrown naked into the tiger cage. Another of Prince Puger's sons, named Raden Sotjo Kusumo, who had tried to stir up a revolt, is being exhibited in a cage on the alu-alu. This took place in the year 1625 of the Javanese calendar, i.e. in the year 1703 of the European calendar).
The story referred to above may be found in a different form in the *Babad Tanah Djawi,* translation by W. L. Olthof, 's-Gravenhage 1941, Vol. II, pp. 262—265. Concerning the maltreatment of the servants, see François Valentijn, *Oud en Nieuw Oost-Indiën,* Vol. IV, p. 146.

Provenance: Presented by dr. Th. Pigeaud, 1963; probably from the collection of W. F. Engelbert van Bevervoorde.

★

H. 699 *Rontèks.*
Water-colours. Probably drawn in Solo, 1888. 30 × 42 cm.
Note: In manuscript H. 699 containing various archive documents on the Principalities of Jogjakarta and Surakarta there is a double sheet with pictures of Rontèks (flags on a short pike used in pagenats, etc.) as used at tigerfights in Surakarta. With Javanese captions.

Provenance: Bequest G. P. Rouffaer, 1928.

★

H. 379 The manuscript "Statistiek van Java, Residentie Grissee" (Statistics of Java, Residency of Gresik), 1822, by A. D. Cornets de Groot, contains a number of ink and sepia drawings, mostly in Javanese style. The work consists of nine forms (A—J) (issued by the Government and filled in bij officials) and memoranda with additional comments. These drawings are found on several sheets added to some of these memoranda.

Provenance: Presented by J. E. Cornets de Groot van Kraayenburg in 1856.

References: Partly published in *T.N.I.*, 1852 & 1853. A copy of the manuscript (the original one?) is in Djakarta.

pp. 333—395 Javanese weapons A—P:

A *Krissen, geheten 1. Biemo Kroedo; 2. Doewong Tjoro Djawie; 3. Doewong Tjoro Balie; 4. Doewong Tjoro Intje* (Krisses, named 1. Bimo Krudo; 2. Duwong Tjoro Djawi; 3. Duwong Tjoro Bali; 4. Duwong Tjoro Intje).

B—H *42 Krislemmetten* (42 Kris-blades).

I—K *Krisscheden en gevesten* (Kris-sheaths and hilts).

L *8 Wapens door de geestelijkheid gebruikt* (8 Weapons used by religious officials).

M—O *21 Piekijzers* (21 Pike-points).

P *10 Pijlen, een boog en een pijlkoker* (10 Arrows, a bow and a quiver).

pp. 397—416 *Zinnebeeldige voorstelling der woekoe's of tijdrekening van 30 weken* (Symbolical representation of the wukus or a 30-week calendar).
Note: Description on pp. 222—232 of the memorandum.

pp. 417—420 *Aantoning van de onderscheidene afbeeldingen der (8) Boedoesche Jaren (voorgesteld door dieren)* (Representation of the various figures by which the 8 years of the "Buddhist" cycle are depicted (represented by animals)).

pp. 421—428 *Wajang Purwa & Wajang gedog.*

pp. 423—460 *Large wedding-procession.*
Note: A—H with 38 figures. Two drawings have been reproduced in *Javaansche Volksvertoningen* by Th. Pigeaud, Batavia 1938, figs. 76—77.

pp. 461—472 *Gamelan.*
Note: A Sléndro. B Pélog. C Senèn & Tjara Bali.

pp. 489—492 *Gereedschap voor de landbouw, natte en droge cultuur* (Farming implements, for wet and dry cultivation).
Note: These drawings follow Form C "Landbouw" (agriculture).

pp. 655—658 *Werktuigen tot het gieten van geschut* (Implements for the casting of cannons).
Note: These drawings follow Forms D—F, "Veeteelt" (cattle-breeding), "Bosschen en Fabrieken" (forests and factories) and "Middelen van Industrie" (methods of industry).

pp. 659—676 *Salpetermakerij* (Salpetre factory).
Note: Water-colours in European style. Description on pp. 648—650 of Memorandum.

pp. 721—728 *Visserij* (Fishing trade).
Note: Description in Memorandum, pp. 688—693.

pp. 729—752 *Verschillende handwerken* (Various crafts).

C *Gold and silversmith-shop.*

D *Brass-foundry and brass-shop.*

E *Forge, gamelan factory.*

F *Carpentry, woodturning etc.*

G *Spinning and batik.*

H *Weaving, hanging and horizontal loom.*
Note: Description of C—H on pp. 693—714 of Memorandum.

pp. 753—788 *Kleedjes* (Textiles).
Note: 108 patterns on 9 sheets (pp. 1—3 "Gebatikte kleedjes" (Textiles with batik patterns); 4—8 "Zijden kleedjes" (Patterns on silk); p. 9 "Garen geweven kleedjes" (Woven patterns)). Commented on and reproduced in *De Batikkunst in Nederlandsch-Indië en hare geschiedenis* by G. P. Rouffaer and H. H. Juynboll, Utrecht 1914.

pp. 965—984 *Hindu-Javanese images.*
Note: These drawings follow Form J "Kunsten en Wetenschappen" (Arts and Sciences). Cf. pp. 909—936.

p. 966 *Sang Yang Temboeroe* (Sang Yang Temburu).

p. 967 *Batoro Goeroe* (Batoro Guru).

p. 970 *Merodjo Maysso Soero* (Merodjo Mayso Suro).

p. 971 *Batoro Romo.*

p. 972 *Ganesja.*

p. 973 *Batoro Gono.*

p. 978 *Batoro Kresno.*

p. 979 *Batoro Siewa.*

p. 982, 983 *Batarie Doergo* (Batari Durgo).

pp. 985—1004 *Reminiscenses of the first advent of Islam (graves).*
Note: A discourse on the bringers of Islam to Java is contained in the Memorandum, pp. 936—949.

pp. 985—988 *[Graf van] Sech Maoelono Magfoer of Malik Ibrahim* (Grave of Shaikh Maulono Magfur or Malik Ibrahim).

pp. 989—992 *[Graf van] Potrie Dewie Swarie* (Grave of Potrie Dewie Swarie).

pp. 993—996 *[Graf van] Njay Gede Pinatih* (Grave of Njay Gede Pinatih).

pp. 997—1000 *[Graf van] Soesoenan Ratoe Girie* (Grave of Susunan Ratu Girie).

pp. 1001—1004 *[Graf van] den Soesoehoenan Prapen* (Grave of Susuhunan Prapen).

pp. 1005—1008 *De kris Soero-Angon-Angon* (The kris Suro-Angon-Angon).

pp. 1009—1020 *Copies of the drawings of gamelans on p. 461—472.*
Note: A discourse on gamelans is contained in the Memorandum, pp. 949—954.

★

F. 16 *Balinese children's drawings.*
In 1939 the Inspector of education Mr. H. te Flierhaar was instructed by the Department of Education and Public Worship in the Neth. Indies, to adapt the system of education in Bali to the Balinese culture in a better way. The results obtained in art classes under the new system in Bali were shown at an exhibition in Djogjakarta on the occasion of the Congress of the Java Institute in the Sana-Budaja Museum (secretary S. Koperberg). At the same time a collection of drawings by Javanese school-children was shown.
The Balinese collection is kept in a special cupboard 59 × 51 × 45 cm with 10 drawers. The pencil drawings are mounted on pasteboard with captions typed on labels, one Dutch on the left and one in English on the right hand side, furnishing the following data: subject, type of school, grade, boy or girl, age. Only a few children signed the drawing with their names. On several sheets there are drawings with notes in Dutch and English beside the general caption. The English of the captions is rather unidiomatic in places. This catalogue therefore does not always offer the exact copies. Sometimes the name of the village where the school is situated is mentioned. These names have not been copied here.

F. 16, 1—98 *Drawings from primary schools.*

F. 16, 1—35 *Decorative flower-motifs.*
Note: F. 16, 1—5: by boys, grade 1, ages 6—8 (5 drawings on one sheet).
F. 16, 6—10: by girls, grade 1, ages 6—8 (5 drawings on one sheet, one is signed Ni Dj. Ngoerah).
F. 16, 11—19: by boys, grade 2, ages 8—10 (9 drawings on 2 sheets).
F. 16, 20—26: by girls, grade 2, ages 8—10 (7 drawings on 2 sheets, one is signed Ni Toeni).
F. 16, 27—30: by boys, grade 3, ages 9—13 (4 drawings on one sheet).
F. 16, 31—35: by girls, grade 3, ages 9—13 (5 drawings on one sheet).

F. 16, 36—66 *Mythological characters and scenes from wayang-plays.*

F. 16, 36 *Radjah* (Magic design).

F. 16, 37 *Sempati* (Bird-ape).
Note: F. 16, 36 and 37: by boys, grade 1, ages 6—8 (2 drawings on one sheet).

F. 16, 38 *Slangenvorst* (King of the Snakes).

F. 16, 39 *Wajang figuur* (Figure from a Wayang-play).

F. 16, 40 *Kinderfantasie* (Child's fantasy).
Note: F. 16, 38—40: by girls, grade 1, ages 6—8 (3 drawings on one sheet).

F. 16, 41 *Gevecht tussen Anoman en Rawana* (Hanuman and Rawana engaged in combat).

F. 16, 42 *Raksasa.*

F. 16, 43 *Held uit het Mahabharata* (Hero from the Mahabharata).

F. 16, 44 *Stenen werpende aap (Ramajana)* (Monkey throwing stones (from the Ramayana)).

F. 16, 45 *Ganeça.*

F. 16, 46 *Anoman en Soegriwa* (Hanuman and Sugriwa).

F. 16, 47—48 *Garoeda* (Garuda).

F. 16, 49 *Ardjoena* (Arjuna).

F. 16, 50 *Apenkoning* (King of the Monkeys).
Note: F. 16, 41—50: by boys, grade 2, ages 8—10 (10 drawings on 3 sheets).

F. 16, 51 *Slangenvorst* (King of the Snakes).

F. 16, 52 *Kresna.*

F. 16, 53 *Toealèn* (Tualèn).
Note: F. 16, 51—53: by girls, grade 2, ages 8—10 (3 drawings on one sheet).

F. 16, 54 *Sang Antaboga.*

F. 16, 55 *Anoman.*

F. 16, 56 *Djatajoe* (Jatayu).

F. 16, 57 *Slangenvorst* (King of the Snakes).

F. 16, 58 *Apenvorst* (King of the Monkeys).

F. 16, 59 *Batara Gana.*

F. 16, 60 *Batara Baroena* (Batara Baruna).

F. 16, 61 *Rama, Laksmana en Djatajoe* (Rama, Laksmana and Jatayu).

F. 16, 62 *Abimanjoe door Djajadrata gedood* (Abimanyu killed by Jayadrata).

F. 16, 63 *Ardjoena, Sangoet, Delem en een priester* (Arjuna, Sangut, Delem and a priest).

F. 16, 64 *Strijd tussen Soebali en Soegriwa* (Subali and Sugriwa engaged in combat).
Note: F. 16, 54—64: by boys, grade 3, ages 9—13 (11 drawings on 6 sheets).

F. 16, 65 *Slangenvorst* (King of the Snakes).

F. 16, 66 *Delem.*
Note: F. 16, 65 and F. 16, 66: by girls, grade 3, ages 9—13 (2 drawings on one sheet).

F. 16, 67—98 *Realistic subjects from daily life.*

F. 16, 67 *Wajang figuren* (Figures from a Wayang-play).

F. 16, 68 *Het karnen van de oceaan* (The churning of the ocean).

F. 16, 69 *Lijkkist* (Coffin).

F. 16, 70 *Barong kèkèt en rangda* (Barong Kèkèt and Rangda).
Note: Representation of a public performance.
Note: F. 16, 67—70: by boys, grade 1, ages 6—8 (4 drawings on one sheet).

F. 16,71 *Lijkverbrandingstoren* (Cremation tower).
Signed Ni Tjanderi.

F. 16, 72 *Barong kèkèt* (Mythical lion).

F. 16, 73 *Baris dansers* (Baris dancers).

F. 16, 74 *Djaoek dansers* (Jauk dancers).
Note: F. 16, 71—74: by girls, grade 1, ages 6—8 (4 drawings on one sheet).

F. 16, 75 *Rangda in een grot* (Rangda in a cave).

F. 16, 76 *Gespleten tempelpoort* (Split temple gate).

F. 16, 77 *Barong kèkèt.*

F. 16, 78 *Stoel waarin de vooroudera*s *naar zee wordt gedragen* (Chair in which the ashes of the ancestors are carried to the sea).

F. 16, 79 *Lijkkist* (Coffin).

F. 16, 80 *Barisdanser* (Baris dancer).

F. 16, 81 *Vooroudertempeltje* (Ancestor shrine).
Note: F. 16, 75—81: by boys, grade 2, ages 8—10 (7 drawings on 2 sheets).

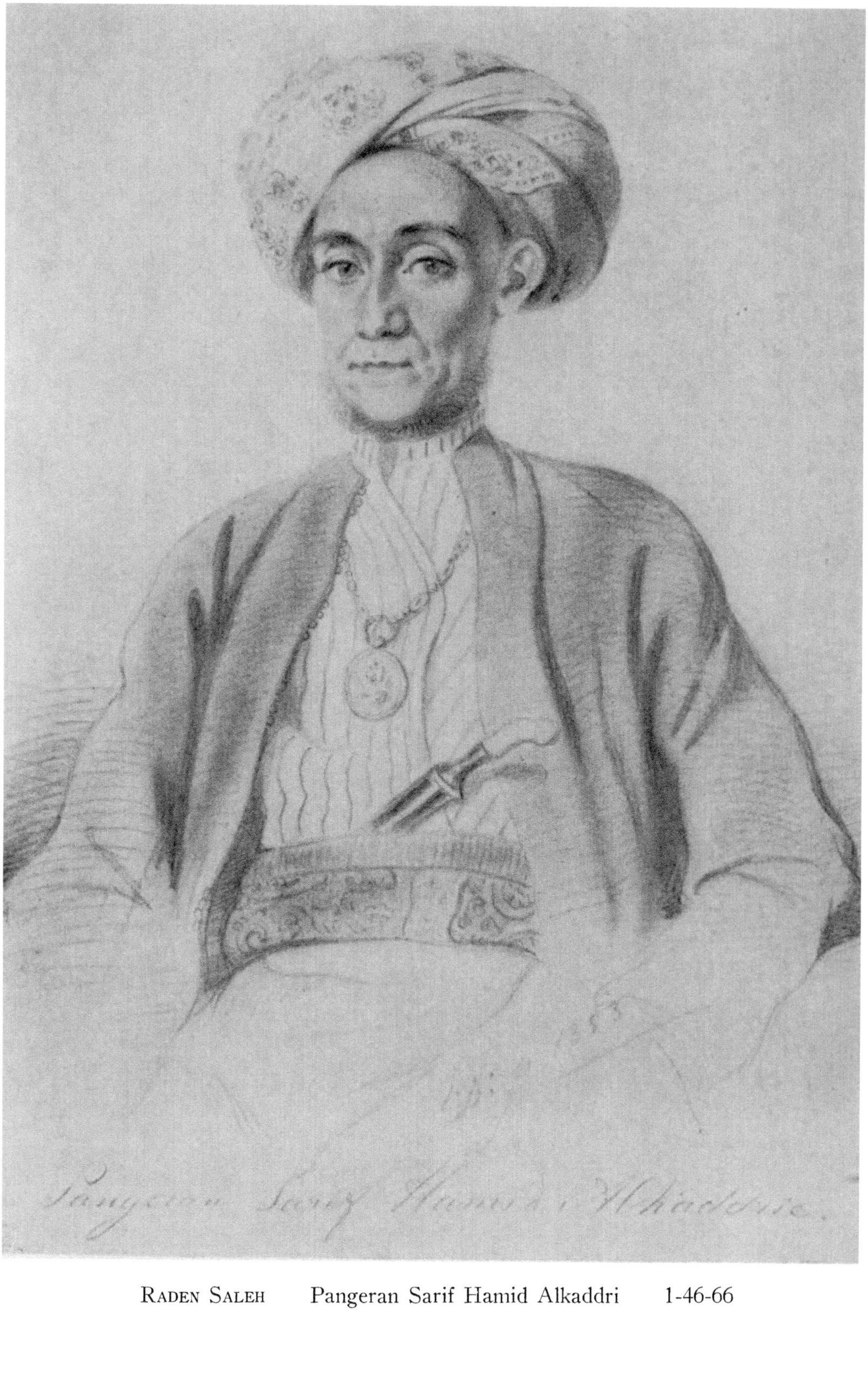

RADEN SALEH   Pangeran Sarif Hamid Alkaddri   1-46-66

F. 16, 82 *Djangèr-danseressen* (Jangèr dancers).

F. 16, 83 *Gespleten poort* (Split gate).

F. 16, 84 *Tempelpoort* (Temple gate).
Note: F. 16, 82—84: by girls, grade 2 ages 8—10 (3 drawings on one sheet).

F. 16, 85 *Tempelpoort* (Temple gate).

F. 16, 86 *Gong met gongstandaard* (Gong with gong-stand).

F. 16, 87—88 *Offerhuisje* (Shrine).

F. 16, 89 *Penassar* (Penasar).

F. 16, 90 *Gespleten tempelpoort* (Split temple-gate).

F. 16, 91 *Lijkkist* (Coffin).
Note: F. 16, 85—91: by boys, grade 3, ages 9—13 (7 drawings on 2 sheets).

F. 16, 92 *Offerhuisje* (Shrine).

F. 16, 93 *Vrouw* (Woman).

F. 16, 94 *Hanengevecht* (Cock-fight).

F. 16, 95 *Scène uit het Tantri-verhaal* (Scene from the Tantri stories).

F. 16, 96 *Lijkverbrandingstoren* (Cremation tower).

F. 16, 97 *Penassar* (Penasar).

F. 16, 98 *Rangda, de Heksenkoningin* (Rangda, Queen of the Witches).
Note: F. 16, 92—98: by girls, grade 3, ages 9—13 (7 drawings on 2 sheets).

F. 16, 99 *Bloemmotief* (Flower-motif).
Ink and sepia. Signed (on reverse side) K. Tinta.
Note: The artist was a student of the primary schoolteachers' course.

F. 16, 100—256 *Drawings from intermediate high schools.*

F. 16, 100—167 *Decorative flower-motifs.*

F. 16, 100—101 *Bloemmotief met Karangmanoek* (Flower-motif with Karangmanuk).

F. 16, 102—103 *Bloemmotief met Karangboma* (Flower-motif with Karangboma).
Note: F. 16, 100—103: by boys, grade 4, ages 10—15 (4 drawings on 4 sheets).

F. 16, 104 *Bloemmotieven* (Flower-motifs).
Note: by girl, grade 4, age 10—15.

F. 16, 105—107 *Bloemmotieven* (Flower-motifs).

F. 16, 108 *Bloemmotief met Karang Boma* (Flower-motif with Karang Boma).

F. 16, 109 *Bloemmotief* (Flower-motif).
Note: Unfinished.

F. 16, 110 *Bloemmotief met Karang Boma* (Flower-motif with Karang Boma).
Ink and sepia.

F. 16, 111—112 *Bloemmotieven* (Flower-motifs).
Ink and sepia.

F. 16, 113—114 *Ganeça-kop met bloemmotief* (Ganeça's head with flower-motif).

F. 16, 115—119 *Bloemmotieven* (Flower-motifs).
Note: F. 16, 119 in ink and sepia.

F. 16, 120 *Ganeça-kop met bloemmotief* (Ganeça's head with flower-motif).

F. 16, 121 *Bloemmotief met Nagaradja* (Flower-motif with the King of the snakes).

F. 16, 122 *Bloemmotief* (Flower-motif).
Ink and sepia.
Note: F. 16, 105—122: by boys, grade 5, ages 12—15 (18 drawings on 15 sheets).

F. 16, 123—125 *Bloemmotieven* (Flower-motifs).
Note: by girls, grade 5, ages 12—15 (3 drawings on one sheet).

F. 16, 126—166 *Flower-motifs.*
Note: 41 drawings on 33 sheets with captions lacking. F. 16, 126, 131 and 133 in ink and sepia, F. 16, 142 in ink. Names of artists: Ng. Sandi, 24th February, 1939; Mad Dadi, 24th February, 1939.

F. 16, 167—224 *Mythological subjects from Wayang-plays.*

F. 16, 167 *Kresna.*
Ink and sepia.

F. 16, 168 *Rawana.*
Ink and sepia.

F. 16, 169 *Batara Gana.*
Ink and sepia.

F. 16, 170—171 *Scène uit het Tantri-verhaal* (Scene from the Tantri stories).
Ink and sepia.

F. 16, 172 *Toealen* (Tualen).

F. 16, 173 *Karang-boma.*

F. 16, 174 *Raksasa.*

F. 16, 175 *Kala Rahoe verslindt de maan* (Kala Rahu swallowing the moon).
Ink and sepia.

F. 16, 176 *Kapi, Djembawan en Ménda* (Kapi, Djembawan and Ménda).

F. 16, 177 *Delem en Koembakarna* (Delem and Kumbakarna).

F. 16, 178 *Twee slangen* (Two snakes).

F. 16, 179 *Rangda, de Heksenkoningin* (Rangda, Queen of the Witches).

F. 16, 180 *Raksasa Geni.*

F. 16, 181 *Djetajoe* (Jetayu).

F. 16, 182 *Anggada.*

F. 16, 183 *Toealèn* (Tualèn).
Note: F. 16, 18—183 on one sheet.

F. 16, 184 *Rama en Sarpanaka* (Rama and Sarpanaka).
Ink and sepia.

F. 16, 185 *Slangenvorst* (King of the Snakes).

F. 16, 186 *Kala Soengsang* (Kala Sungsang).
Ink and sepia.

F. 16, 187 *Anoman.*
Ink and sepia.

F. 16, 188 *Ardjoena in gevecht* (Arjuna engaged in combat).

F. 16, 189 *Barong Kèkèt.*

F. 16, 190 *Garoeda* (Garuda).
Note: F. 16, 167—190: by boys, grade 4, ages 10—15 (24 drawings on 20 sheets).

F. 16, 191 *Hellewachter* (Doorkeeper of Hell).

F. 16, 192 *Demon.*

F. 16, 193 *Kala Soengsang* (Kala Sungsang).

F. 16, 194 *Anoman.*

F. 16, 195 *Sarpanaka.*

F. 16, 196 *Antaboga.*

F. 16, 197 *Kresna.*
Note: F. 16, 195—197 on one sheet.

F. 16, 198 *Deurwachter van de hel* (Doorkeeper of Hell).

F. 16, 199 *Wajang figuur* (Wayang figure).
Note: F. 16, 198 and F. 16, 199 on one sheet.

F. 16, 200 *Rama, Leksmana en het apenleger* (Rama, Leksmana and the army of the monkeys).

F. 16, 201 *Koembakarna door de apen aangevallen* (Kumbakarna attacked by the monkeys).

F. 16, 202 *Anoman en Soegriwa* (Hanuman and Sugriwa).

F. 16, 203 *Strijd van het apenleger tegen de patih van Langka* (Battle between the monkeys and the prime-minister of Langka).

F. 16, 204 *Swadana leerling van Rangda* (Swadana, pupil of Rangda).

F. 16, 205 *Demon.*

F. 16, 206 *Rama en Djatajoe* (Rama and Jatayu).

F. 16, 207 *Legeraanvoerder van het apenleger* (General of the monkey army).

F. 16, 208 *Soebali en Sang Rawana* (Subali and Sang Rawana).
Note: These names appear on the drawing in Balinese script.

F. 16, 209 *Sangoet en Delem* (Sangut and Delem).

F. 16, 210 *Sang Wimana.*

F. 16, 211 *Merdah.*
Note: F. 16,208—211 on one sheet.

F. 16, 212 *Anoman en Toealèn* (Hanuman and Tualèn).

F. 16, 213 *Kalaraoe* (Kalarahu).

F. 16, 214 *Sang Rawana.*

F. 16, 215 *Mythische aap* (Mythical monkey).
Note: F. 16, 212—215 on one sheet.

F. 16, 216 *Strijd tussen Koembakarna en de apen* (Battle between Kumbakarna and the monkeys).

F. 16, 217 *Gevecht tussen Rangda en Pandoeng* (Rangda and Pandung engaged in combat).

F. 16, 218 *Gevecht tussen Anoman en Rawana* (Hanuman and Rawana engaged in combat).

F. 16, 219 *Sterrenbeeld Koemba (watervat)* (The constellation of Kumba (water-jar)).

F. 16, 220 *Rangda, de Heksenkoningin* (Rangda, Queen of the Witches).

F. 16, 221 *Rama.*
In colours.

Note: F. 16,191—221: by boys, grade 5, ages 12—15 (31 drawings on 17 sheets).

F. 16, 222 *Prinses met Toealèn en Merdah* (Princess with Tualèn and Merdah).
Note: By girl, grade 5, ages 12—15.

F. 16, 223 *Onderwerpen uit de wajang-geschiedenis* (Characters and scenes from Wayang stories).

F. 16, 224 *De reus en de aap* (The giant and the monkey).
Note: F. 16, 223 and F. 16, 224 drawings without information as to grade, etc.

F. 16, 225—256 *Realistic subjects from daily life.*

F. 16, 225 *Hanengevecht* (Cock-fight).

F. 16, 226 *Gamelan instrument.*
Signed K. Poetera.

F. 16, 227 *Draagstoel* (Sedan chair).
Signed I Da Pt Nastre.

F. 16, 228 *Draagstoel* (Sedan chair).
Signed I. Jasa.
Note: F. 16, 225—228 on one sheet.

F. 16, 229 *Lijkverbrandingstoren* (Cremation tower).

F. 16, 230 *Tor* (Beetle).
Ink and pencil.

F. 16, 231 *Lijkkist* (Coffin).
Note: F. 16, 225—231: by boys, grade 4, ages 12—15 (7 drawings on 4 sheets).

F. 16, 232 *Vrouw* (Woman).

F. 16, 233 *Offerhuisje* (Shrine).
Signed Dajoe Md. Tirtawati.
Note: F. 16, 232 and F. 16, 233: by girls, grade 4, ages 12—15.

F. 16, 234—235 *Rangda de Heksenkoningin* (Rangda, Queen of the Witches).

F. 16, 236 *Maansverduistering; Kala Rahoe verslindt de maan* (An eclipse of the moon; Kala Rahu swallowing the moon).
Note: F. 16,234—236 on one sheet.

F. 16, 237 *Scène uit het verhaal van Tjoepak en Grantang* (Scene from the story of Tjupak and Grantang).

F. 16, 238 *Landschap* (Landscape).

F. 16, 239 *Gevecht tussen olifant en leeuw* (Fight between an elephant and a lion).
Note: F. 16, 237—239 on one sheet.

F. 16, 240—241 *Gong met standaard* (Gong with stand).

F. 16, 242 *Tempelpoort* (Temple gate).

F. 16, 243 *Lijkverbrandingsprocessie* (Cremation procession).

F. 16, 244 *Scène uit het dagelijks leven* (Scene from everyday life).

F. 16, 245 *Landschap* (Landscape).

F. 16, 246 *Godsdienstige optocht* (Religious procession).

F. 16, 247 *Barong Kèkèt.*
Ink and sepia.

F. 16, 248 *Kala Rahoe verslindt de maan* (Kala Rahu swallowing the moon).

F. 16, 249 *Masker* (Mask).

F. 16, 250 *Gamelan orkest voor de tempelpoort* (Gamelan orchestra in front of temple gate).

F. 16, 251 *Twee vogels* (Two birds).

F. 16, 252 *Landschap* (Landscape).

F. 16, 253 *Clownesque figuur* (Clown-like figure).
Note: F. 16, 234—253: by boys, grade 5, ages 12—15 (20 drawings on 16 sheets).

F. 16, 254 *Meroe* (Meru).

F. 16, 255 *Tjandi bentar* (Split (temple) gate).

F. 16, 256 *Offerhuis* (Shrine).
Note: F. 16, 254—256: by girls, grade 5, ages 12—15.

F. 16, 257—278 *Unmounted drawings, mostly with captions on reverse side.*

F. 16, 257 *Standing figure with elephant's mask, raising a sword.*
Ink and sepia. Signed Made Sajang, 18 th., (= aged 18 years), 24-6-39. 37 × 32 cm.

F. 16, 258 *Arjuna, shooting with an arrow.*
Pencil. Signed Dw. Pt. Bading, 18 th., 24-6-39. 37 × 32 cm.

F. 16, 259 *Naga.*
Pencil. Signed Wj. Soedana, 19 th., 24-6-39. 39 × 32 cm.
Note: F. 16, 257—259: by male pupils of the Cursus Volks Onderwijs (Primary schoolteachers' course), 2nd class, Den Passar.

F. 16, 260 *Gong met standaard* (Gong with stand).
Ink and sepia. 39 × 32 cm.
Note: Typed caption: Cursus Volks Onderwijs, Kloengkoeng, jongen P. 5 (Primary schoolteachers' course, Klungkung, boy P. 5).

F. 16, 261 *Naga dengan Paksi* (Naga and Garuda).

Pencil. Signed I. Rènoh, dessaschool Kediri (Tabanan), 1/10 '40. 37 × 32 cm.
Note: With caption: "Dikerdjakan diroemah" (drawn at home).

F. 16, 262 *Soebali.*
Pencil. Signed I. Gst. Md. Lipoer, Belajoe. 37 × 32 cm.
Note: Subali, King of the monkeys with "panakawans" (servants) and Sangut.

F. 16, 263 *Bima [and] Rangda.*
Pencil. Signed Ida Bgs. Oka, Belajoe. 37 × 32 cm.

F. 16, 264 *Watekewatja [and] Sangoet.*
Pencil. Signed I. Gst. Md. Lipoer, Belajoe. 37 × 32 cm.

F. 16, 265 *Kombakarna berperang dengan Soegeriwa* (Fight between Kumbakarna and Sugriwa).
Pencil. Signed Njoman Rai, di Kediri (Tabanan). Dated 18 Januari 1941. 37 × 32 cm.
Note: With caption: "Dikerdjakan diroemah" (drawn at home).

F. 16, 266 *Kombakarna berperang dengan Anoman* (Fight between Kumbakarna and Anoman).
Pencil. Signed Wajan Rènoh, di Kediri (Tabanan). Dated 18 Januari 1941. 37 × 32 cm.
Note: With caption: "Dikerdjakan diroemah" (drawn at home).

F. 16, 267 *Sjiwa and Arjuna.*
Pencil. Signed I. Gebioeh, moerid sek. desa Satrija (pupil of the Satrija village-school), Den Passar. 37 × 32 cm.

F. 16, 268 *Celestial nymphs trying to distract Arjuna who is practising ascetism.*
Pencil. Signed I. Gebioeh, sek. desa Satrija (Satrya village-school), Den Passar. 37 × 32 cm.
Note: F. 16, 261—268: by male pupils of a village-school, 3rd class.

F. 16, 269 *Boar eating from a trough, sow and piglets.*
Ink and sepia. Signed I. Sukadi. 25 × 35 cm.

F. 16, 270 *Scene from a story.*
Ink and sepia. Signed I. Rekuk. 25 × 35 cm.
Note: The drawing shows a woman with a fan in her hand running away from a kneeling youth. A dressed-up dog (?) is holding an umbrella.
Note: F. 16, 269 and F. 16, 270: signed on reverse side in Balinese script. These are not school-drawings.

F. 16, 271 *Dancer.*
Ink and water-colours. 37 × 28 cm.

F. 16, 272—278 *Decorative pencil drawings.*
Note: Made by boys of "sek. vervolg di Kediri" (intermediate high school of Kediri).

F. 16, 272 *Decorative drawing.*
Signed I. Kt. Tama, kl. IV (4th class). Dated 30-8-40. 12,5 × 18,5 cm.

F. 16, 273 *Decorative drawing.*
Signed I. Goestri Djam, kl. IV (4th class). Dated 27-12-40. 12,5 × 18,5 cm.

F. 16, 274 *Decorative drawing.*
Signed I. Kt. Tama, kl. IV (4th class). Dated 13-9-40. 12,5 × 18,5 cm.

F. 16, 275 *Decorative drawing.*
Signed I. Goestri Djam, kl. IV (4th class). Dated 4-10-40. 12,5 × 18,5 cm.

F. 16, 276 *Decorative drawing.*
Signed Gs. Pt. Pogog, kl. IV (4th class). Dated 30-8-40. 12,5 × 18,5 cm.

F. 16, 277 *Decorative drawing.*
Signed I. Ribong, kl. II (2nd class). 18,5 × 12,5 cm.

F. 16, 278 *Decorative drawing.*
Signed Gs. Nj. Rapit, kl. IV (4th class). Dated 30-8-40. 12,5 × 18,5 cm.

F. 16, 279—281 *Menu cards.*
Ink and coloured chalk. 12,5 × 8,5 cm.
Note: Three menus used at a luncheon to celebrate the 10th anniversary of the Siladarma association of parents and friends, on Monday 4th July, 1938 at Klungkung.

References: H. te Flierhaar, "De aanpassing van het Inlandsch onderwijs op Bali aan de eigen sfeer", *Koloniale Studiën,* 1941, pp. 135—159; and "Hervorming van het teekenonderwijs", pp. 143—147 (illustrated).

★

Ornamented letters

Letters sent to native chiefs by governor generals and other high officials, and vice versa, were often decorated with coloured and gilded ornamental borders. The collection of Oriental manuscripts contains the following examples of such letters.

Cod. Or. 121 *Letter from the Sultan of Bima to Baron van der Capellen.*
45 × 32,5 cm. Border in water-colours, blue and pink with gold.

Cod. Or. 122 *Two letters, one from the Sultan of Sumenep and one from the Panembahan of Madura, to the Resident of Surabaya.*
40 × 27 cm and 44 × 35 cm. Dated '1 Jan. 1854'. Border in water-colours with naturalistic representations of roses and leaves.

Cod. Or. 123 *Letter from the Sultan of Sumenep to Baron van der Capellen.*
52 × 41 cm. Dated 27 Feb. 1826. Ornamental border in gold and black water-colours with design of vine tendrils.

References: Ph. S. van Ronkel, "Catalogus der Maleische handschriften" *B.K.I.* LX, 1908, p. 235, Nrs. LXXIV, LXXV and LXXVI.

Cod. Or. 172 *Collection of 52 letters from governor generals, etc., to the Panembahan, later Sultan, of Sumenep.*

1 *Letter from Sir Th. S. Raffles.*
56 × 48 cm. Dated 17 Nov. 1815.
Note: Left part in Arabic characters, with the English text on the right. Both parts decorated with an ornamental border. At the top is the English Coat of Arms.

Reference: Ph. S. van Ronkel, "Een Maleische afscheidsbrief van Raffles", *B.K.I.* CIII, 1946, pp. 459—468.

2 *Letter from L. C. Graaf van Ranzow at Sumenep.*
52,5 × 40 cm. Dated '22 Juni 1824'. Ornamental border with crown, design of flowers and leaves in gold.

3—52 *Letters from Dutch governor generals, etc.*
45,5 × 37 cm. Ornamental borders, evidently based on one pattern but differing in details of design and colouring. Gold is always used.

Cod. Or. 173 *Two letters from the Panembahan of Pamekasan.*

1 *Addressed to H. Cornets de Groot, assistant resident of Sumenep and Pamekasan.*
40 × 25 cm. Dated Pamekasan 11 Aug. 1830. Border in water-colours with gold, Louis XVI ornamentation.

2 *Addressed to H. Cornets de Groot, resident of Besuki.*
43 × 27 cm. Dated Sumenep 1 Jan. 1840. Border in water-colours with gold, Louis XV ornamentation.

References: Ph. S. van Ronkel, "Aanvulling der beschrijving van de Maleische en Miningkabausche handschriften", *B.K.I.* CIII, 1946, pp. 580—591, Nrs. CXLIX and CL.

INDEX OF NAMES

T

KONINKLIJK INSTITUUT

VOOR TAAL-, LAND- EN VOLKENKUNDE

CATALOGUES

Maps:

Catalogus der Land- en Zeekaarten toebehoorende aan het Koninklijk Instituut voor de Taal-, Land- en Volkenkunde van Nederlandsch-Indië samengesteld door G. P. Rouffaer, 's-Gravenhage 1898. 150 pp. *Out of print.*

Catalogus der Land- en Zeekaarten van- en uitgegeven door het Koninklijk Instituut voor de Taal-, Land- en Volkenkunde van Ned.-Indië samengesteld door W. C. Muller, 's-Gravenhage 1913. 150 pp. *Out of print.*

Books and periodicals:

Catalogus der Koloniale Bibliotheek van het Koninklijk Instituut voor de Taal-, Land- en Volkenkunde van Ned.-Indië en het Indisch Genootschap door G. P. Rouffaer en W. C. Muller. 's-Gravenhage 1908. 1053 pp.

Idem. Eerste Supplement door W. C. Muller. 's-Gravenhage 1915. 426 pp.

Idem. Tweede Supplement door W. C. Muller. 's-Gravenhage 1927. 458 pp.

Idem. Derde Supplement door W. J. M. Buch. 's-Gravenhage 1937. 438 pp.

Idem. Vierde Supplement. 's-Gravenhage 1966. 801 pp.

Western Manuscripts:

Catalogus van de handschriften in Westerse talen, toebehorende aan het Koninklijk Instituut voor Taal-, Land- en Volkenkunde door H. J. de Graaf. 's-Gravenhage 1963. 184 pp.

Eastern Manuscripts:

Catalogus der Maleische handschriften van het Koninklijk Instituut voor de Taal-, Land- en Volkenkunde van Ned.-Indië door Dr. Ph. S. van Ronkel. *Bijdragen tot de taal-, land- en volkenkunde van Ned.-Indië* part 60 (1908) pp. 181-248.

Catalogus der Javaansche, Balineesche en Madureesche handschriften van het Koninklijk Instituut voor de Taal-, Land- en Volkenkunde van Ned.-Indië door Dr. H. H. Juynboll. *Bijdragen tot de taal-, land- en volkenkunde van Ned.-Indië* part 69 (1914) pp. 386-418.

Aanvulling der beschrijving van de Maleische en Minangkabausche handschriften, benevens een Atjèhsch handschrift, in het bezit van het Koninklijk Instituut voor de Taal-, Land- en Volkenkunde van Ned.-Indië door Dr. Ph. S. van Ronkel, *Bijdragen tot de taal-, land- en volkenkunde van Ned.-Indië* part 103 (1946) pp. 555-606.